The Welfare of Cats

Edited by

Irene Rochlitz

*University of Cambridge,
Cambridge, United Kingdom*

 Springer

A C.I.P. Catalogue record for this book is available from the Library of Congress.

ISBN 978-1-4020-6143-1 (PB)
ISBN 978-1-4020-3226-4 (HB)
ISBN 978-1-4020-3227-1 (e-book)

Published by Springer,
P.O. Box 17, 3300 AA Dordrecht, The Netherlands.

Printed on acid-free paper

All Rights Reserved
© 2007 Springer
No part of this work may be reproduced, stored in a retrieval system, or transmitted
in any form or by any means, electronic, mechanical, photocopying, microfilming, recording
or otherwise, without written permission from the Publisher, with the exception
of any material supplied specifically for the purpose of being entered
and executed on a computer system, for exclusive use by the purchaser of the work.

BISHOP BURTON LRC
WITHDRAWN

T052182

636.8

Animal Welfare

VOLUME 3

Series Editor

Clive Phillips, *Professor of Animal Welfare, Centre for Animal Welfare and Ethics, School of Veterinary Science, University of Queensland, Australia*

Titles published in this series:

Volume 1: The Welfare of Horses
 Natalie Waran
 ISBN 1-4020-0766-3

Volume 2: The Welfare of Laboratory Animals
 Eila Kaliste
 ISBN 1-4020-2270-0

TABLE OF CONTENTS

Series Preface	vii
Preface and Acknowledgements	xi
List of Contributors	xvii
Chapter 1 CAT BEHAVIOUR: SOCIAL ORGANIZATION, COMMUNICATION AND DEVELOPMENT By Sharon L. Crowell-Davis	1
Chapter 2 THE ASSESSMENT OF WELFARE By Rachel A. Casey and John W. S. Bradshaw	23
Chapter 3 THE HUMAN-CAT RELATIONSHIP By Penny L. Bernstein	47
Chapter 4 BEHAVIOUR PROBLEMS AND WELFARE By Sarah E. Heath	91
Chapter 5 CAT OVERPOPULATION IN THE UNITED STATES By Philip H. Kass	119

TABLE OF CONTENTS

Chapter 6 — 141
THE WELFARE OF FERAL CATS
By Margaret R. Slater

Chapter 7 — 177
HOUSING AND WELFARE
By Irene Rochlitz

Chapter 8 — 205
DISEASE AND WELFARE
By Kit Sturgess

Chapter 9 — 227
NUTRITION AND WELFARE
By Kit Sturgess and Karyl J. Hurley

Chapter 10 — 259
BREEDING AND WELFARE
By Andreas Steiger

INDEX — 277

ANIMAL WELFARE BY SPECIES: SERIES PREFACE

Animal welfare is attracting increasing interest worldwide, but particularly from those in developed countries, who now have the knowledge and resources to be able to offer the best management systems for their farm animals, as well as potentially being able to offer plentiful resources for companion, zoo and laboratory animals. The increased attention given to farm animal welfare in the West derives largely from the fact that the relentless pursuit of financial reward and efficiency has led to the development of intensive animal production systems, that challenge the conscience of many consumers in those countries. In developing countries human survival is still a daily uncertainty, so that provision for animal welfare has to be balanced against human welfare. Welfare is usually provided for only if it supports the output of the animal, be it food, work, clothing, sport or companionship. In reality, there are resources for all if they are properly husbanded in both developing and developed countries. The inequitable division of the world's riches creates physical and psychological poverty for humans and animals alike in all sectors of the world. Livestock are the world's biggest land user (FAO, 2002) and the population is increasing rapidly to meet the need of an expanding human population. Populations of farm animals managed by humans are therefore increasing worldwide, and there is the tendency to allocate fewer resources to each animal.

Increased attention to welfare issues is just as evident for companion, laboratory, wild and zoo animals. Although the economics of welfare provision may be less critical than for farm animals, the key issues of provision of adequate food, water, a suitable environment, companionship and health remain as important as they are for farm animals. Of increasing importance is the ethical management of breeding programmes, now that genetic manipulation is more feasible, but there is less tolerance of deliberate breeding of animals with genetic abnormalities. However, the quest for producing novel genotypes has fascinated breeders for centuries, and where dog and cat breeders produced a variety of extreme forms with adverse effects on their welfare in earlier times, nowadays the quest is pursued in the laboratory, where the mouse is genetically manipulated with even more dramatic effects.

The intimate connection between animal and owner or manager that was so essential in the past is rare nowadays, having been superseded by technologically efficient production systems, where animals on farms and in laboratories are tended by fewer and fewer humans in the drive to enhance labour efficiency. In today's busy lifestyle pets too may suffer from reduced contact with humans, although their value in providing companionship, particularly for certain groups such as the elderly, is increasingly recognised. Consumers also rarely have any contact with the animals that produce their food. In this estranged, efficient world man struggles to find the moral imperatives to determine the level of welfare that he should afford to animals within his charge. Some, such as many pet owners, aim for what they believe to be the highest levels of welfare provision, while others, deliberately or through ignorance, keep animals in impoverished conditions or even dangerously close to death. Religious beliefs and directives encouraging us to care for animals have been cast aside in an act of supreme human self-confidence, stemming largely from the accelerating pace of scientific development. Instead, today's moral codes are derived as much from media reports of animal abuse and the assurances that we receive from supermarkets, that animals used for their products have not suffered in any way. The young were always exhorted to be kind to animals through exposure to fables, whose moral message was the benevolent treatment of animals. Such messages are today enlivened by the powerful images of modern technology, but essentially still alert children to the wrongs associated with animal abuse.

This series has been designed to provide academic texts discussing the provision for the welfare of the major animal species that are managed and cared for by humans. They are not detailed blueprints for the management of each species, rather they describe and consider the major welfare concerns of the species, often in relation to the wild progenitors of the managed animals. Welfare is considered in relation to the animal's needs, concentrating on nutrition, behaviour, reproduction and the physical and social environment. Economic effects of animal welfare provision are also considered where relevant, and key areas requiring further research.

With the growing pace of knowledge in this new area of research, it is hoped that this series will provide a timely and much-needed set of texts for researchers, lecturers, practitioners, and students. My thanks are particularly due to the publishers for their support, and to the authors and editors for their hard work in producing the texts on time and in good order.

Clive Phillips, Series Editor
Professor of Animal Welfare and Director, Centre for Animal Welfare and Ethics, School of Veterinary Science, University of Queensland, Australia.

Reference:
Food and Agriculture Organisation (2002). http://www.fao.org/ag/aga/index_en.htm.

PREFACE AND ACKNOWLEDGEMENTS

The last decade has seen the publication of several books on animal welfare, companion animal behaviour and behaviour therapy, as well as the relationship between domestic animals and humans. The distinguishing characteristic of this work is that it focuses on the major issues directly affecting the welfare of domestic cats. I hope that this volume will help researchers, animal welfare organisations, cat owners and all those concerned with feline welfare to develop a better understanding of these issues, and provide some guidance as to the ways in which they can be addressed.

An appreciation of feline behaviour is essential in order to identify and tackle welfare problems successfully, so the social organization, methods of communication and development of cat behaviour are considered in Chapter 1. The traditional view of the asocial cat that walks alone no longer holds. It is now clear that the feline social system is flexible, with a complex range of social behaviours that allow cats to live alone or in groups of varying size. Affiliative behaviours, such as touching noses, allogrooming and allorubbing, play and resting together, are described, as well as the other ways in which cats communicate with one another and with humans. Factors affecting the development of behaviour in kittens, socialization and behaviours directed towards humans are discussed, as is the complex topic of feline social hierarchies.

In Chapter 2, methods of assessing welfare in cats are described. The authors adopt the view that feelings and emotions, a reflection of the cat's mental state, determine its welfare. As it is not currently possible to directly probe the mental state of cats, two indirect approaches are proposed. The first examines the attributes of an individual, such as its behaviour and physiological state, seeking to evaluate the extent to which it is coping with its environment. The second approach compares the environment in which the species, or its ancestor, has evolved and to which it is adapted, with its current environment and makes predictions about its likely level of welfare. The difficulties with both these approaches, and how the different welfare measures should be integrated and interpreted, are discussed. As the authors point out, much of the research on the assessment of welfare has been conducted on cats in shelters, catteries and laboratories, with little work on the welfare of strays, cats in feral colonies and pet cats.

The relationship between cats and humans is an important one, and is considered in Chapter 3. Recent findings suggest that cats may have been associated with humans as long as 9,500 years ago (Vigne *et al.* 2004). Currently, there are an estimated 76 million pet cats in the United States, 7.7 million in the United Kingdom and over 200 million worldwide. The benefits that humans and their animal companions may provide for one another are described. Studies examining influences on the socialization of cats to humans are summarized, as are those on cat personality, breed differences, and the few studies examining how cats interact with one another and with humans in the home. The relationship between humans and cats can sometimes fail, and when it does the consequences for the welfare of the cats, and sometimes of their owners too, may be severe. Recent research on animal abuse, animal hoarding, and the connection between animal abuse and violence among humans is presented, as is information on the rapidly developing field of animal law, and efforts to improve the way animals are regarded and treated.

As cats have become more popular, behaviour problems, which are considered in Chapter 4, are reported more frequently. Behaviour problems can result in cats being abandoned, relinquished to shelters, or presented for euthanasia. As the cat's role has changed from that of rodent controller to cherished companion animal, so too have the expectations of its owners. Some normal feline behaviours, such as scratching, predatory-related and general activity, are regarded as problematic largely due to the owner's lack of understanding of normal behaviour, and failure to acknowledge the constraints placed upon it by the domestic environment. The motivations underlying these behaviours and approaches to treatment are discussed. Influences during the early development of cats, on their socialization to other cats and to humans, are all important in preventing future behaviour problems. Because of the way feline social groups are normally formed and organized, the tendency of owners to keep several, often unrelated, cats in a household does not always lead to harmonious relationships. It is increasingly recognised that cats kept in socially incompatible groups may experience chronic stress and poor welfare, and methods of addressing these issues are discussed. Finally, the close relationship between behaviour and disease, and its effect on welfare, are described.

The existence of hierarchies, where there is a fairly well defined social ranking with dominant and subordinate cats, is still being debated among those studying cat social behaviour. In Chapter 1 hierarchical relationships between cats, that involve dominance and submission, are described. Within a group, there may be higher-ranking, or dominant, and lower-ranking, or subordinate, cats, and the higher-ranking cats control the important resources. In Chapters 2 and 4 the alternative opinion, that a specific hierarchy does not form within an established group of cats, is presented. This is an area of cat behaviour that warrants more research, particularly because of the role of social stress in the development of behaviour problems, and the general effects of chronic social stress, tension and conflict on welfare. Whether they are kept in catteries, shelters, laboratories, or in the home, an understanding of the way feline social groups are organized underpins the way we manage and care for them.

One of the most serious issues affecting the welfare of cats, at least in the United States and the United Kingdom, is that of overpopulation, reflected in the ever-increasing numbers of largely healthy cats relinquished to animal shelters. The problem of overpopulation in the United States is considered in Chapter 5. For a variety of reasons, the number of cats entering American shelters, and the number of healthy cats euthanized there, are difficult to estimate. It is clear, though, that these numbers are shockingly high. Understanding the characteristics of the relinquished cats, the reasons why they are relinquished, and the characteristics of their owners, is essential in order to develop methods to address the problem. The findings of a number of surveys on relinquishment, including the Regional Shelter Relinquishment Survey Study, which included 1,409 owners who relinquished cats or their litters to shelters, are presented. In view of the considerable diversity of people and regions across the United States, it is dangerous to make broad generalizations but the common findings shared by these studies are presented. The small amount of research that has been published on the factors that affect the likelihood of cats being adopted from shelters is summarized.

The welfare of the feral cat population is considered in Chapter 6. Again, it is difficult to determine the number of free-roaming or feral cats; the author estimates that the number of feral cats in the United States is about one third to one half the number of owned cats. Methods of population control for feral cats are presented, with trap, neuter and return (TNR) programs being one of the most effective and humane. A range of TNR programs, based in the United States and elsewhere, are described. Follow-up studies, to determine their effectiveness, have been carried out for some of these programs. Ways to reduce the number of cats entering the feral population include ensuring the identification of cats so strays can be returned to their owners, ensuring cats allowed outside are sterilized, and helping owners to deal with their cats' behaviour problems and thereby avoid their abandonment or relinquishment. The emotive issue of predation of wildlife by feral cats is discussed, as are public health issues (such as zoonotic disease), and, finally, the health status and welfare of feral cats themselves.

The way a cat is housed will have a significant impact on its welfare, so the requirements of cats housed in a variety of conditions are described in Chapter 7. The main housing conditions include research facilities; boarding, breeding or quarantine catteries; shelters and sanctuaries; veterinary practices and the home environment. Most of the studies on housing have been conducted in research facilities, catteries, and shelters, but their findings can be applied to other situations. General recommendations are made, with regard to the quantity of space cats need, the quality of the space (that is, what its internal features should provide) and the need of cats to have contact with other cats and with humans. In addition, the requirements of cats to live in a stimulating sensory environment, to have opportunities to explore and play, and to have appropriate access to food and water, are described. Particular considerations for cats kept in research facilities, in shelters, and in the home environment are mentioned. For pet cats in the home, the advantages and disadvantages of keeping them permanently indoors or allowing them outdoor access are discussed.

In Chapter 8, how disease affects welfare is considered. Infectious disease, in particular, is common in cats so principles of infectious disease prevention, such as vaccination, screening for infectious agents and reduction of exposure, are described. Infectious disease may be spread horizontally or vertically, and carrier cats are of great epidemiological importance in disease spread. Methods to control infectious disease include attention to hygiene, the reduction of stress factors, and quarantine and isolation of animals. Ways to prevent and control infectious disease in high-risk groups, such as multi-cat households, catteries, and shelters are considered. Of non-infectious diseases, dental disease, trauma, chronic renal failure and gastrointestinal disease most commonly affect cats. Methods of screening for non-infectious disease, as well as the skills of good history taking and physical examination of feline patients, are described. Finally, the recognition and treatment of pain, an important welfare issue in all species, is considered. Chronic pain is more difficult to identify than acute pain in cats, and is likely to have a more significant impact on welfare than is currently recognised.

The crucial role that nutrition has in determining the health and welfare of cats is described in Chapter 9. In most Western countries, cats no longer rely on wildlife populations for their nutrition but on their owners, who must provide them with nutritionally complete and safe foods, offered in a way that complies with their natural feeding behaviour and physiology. Being obligate carnivores, cats have a much narrower range of tolerance for various dietary components than humans or dogs, and nutritional deficiencies and toxicity problems are relatively more common. Their particular nutritional requirements are listed, as are some of the common problems associated with food. Aspects of feeding behaviour, food selection and patterns of food intake, and the nutritional principles to ensure the optimal health of cats at different life stages (kitten, adult, breeding queen and geriatric) are described. Major advancements have been made in the nutritional management of disease in cats, and general approaches to the management of obesity (see also Chapter 4), and to feeding sick cats, are presented.

In the final chapter, how breeding for extreme characteristics in cats can have adverse effects on their welfare is described. Compared with dogs and many other domestic animals, pedigree cat breeds were developed relatively recently. However, although the majority of domestic cats today are non-pedigree, there is increasing interest in developing new breeds and in modifying existing breeds. As stated in the "European Convention for the Protection of Pet Animals" (Council of Europe 1987), the aim of responsible breeding should be that both parents and offspring are able to live a healthy life, and be capable of their normal species-specific behaviour; considerable deviations in breed morphology, physiology and behaviour may cause pain and suffering. Recommendations on how to interpret and apply the general rules of the Convention, elaborated in the "Resolution on the Breeding of Pet Animals" (Council of Europe 1995), are presented. This chapter includes a table showing the characteristics in cats that may be associated with welfare problems, the underlying genetics (where known) and breeds affected, and the measures that should be taken to improve welfare. The author proposes that detailed morphological, physiological and behavioural criteria, based on welfare considerations, should be defined and used in the assessment of existing and future breeds. However, it may be difficult to reach a consensus on these criteria among such diverse groups as veterinarians, breeders, show judges and owners.

I would like to thank Professor Clive Phillips of the University of Queensland, and Dr. Cristina Alves dos Santos of Springer, for giving me the opportunity to edit this book on "The Welfare of Cats". Cats are fascinating, mysterious and intriguing, and I am glad to be able to learn more, and wonder more, about them. I hope that, in some way, this book will help to improve their welfare and to stimulate research on how to do so even better.

PREFACE AND ACKNOWLEDGEMENTS

I am very grateful to all the authors, who generously contributed their time and efforts without payment in order to bring this important volume to fruition. Their interest in cats, and concern for animal welfare, are evident throughout their work for this book. I would like to thank Melania Ruiz, of Springer, for helping me with the formatting of the chapters and preparing them for publication, and Professor Phillips for carrying out the final edit.

Many people helped me with this project. I am grateful to Professor Don Broom for allowing me to use the facilities at the University of Cambridge, and also to Joy Archer, David Gouldstone, Kristin Hagen, Mark Holmes, Ildiko Plaganyi, Joseph Rochlitz, Michael Rochlitz, Cerian Webb and Jessica Daisy for reviewing chapters, suggesting improvements and listening. Most of all, I would like to thank my family for their encouragement and support throughout.

Irene Rochlitz

References

Council of Europe (1987) European Convention for the protection of pet animals, 13th November 1987 (ETS 125), Council of Europe, F 67075 Strasbourg-Cedex.

Council of Europe (1995) Resolution on the breeding of pet animals, Multilateral Consultation of parties to the European Convention for the protection of pet animals (ETS 123), March 1995 in Strasbourg, Document CONS 125(95)29, Council of Europe, F 67075 Strasbourg-Cedex.

Vigne, J-D., Guilaine, J., Debue, K., Haye, L. and Gerard, P. (2004) Early taming of the cat in Cyprus. *Science* **304**, 259.

LIST OF CONTRIBUTORS

Penny L. Bernstein, PhD
Associate Professor, Biological Sciences, Kent State University Stark Campus, 6000 Frank Avenue, Canton, OH 44720 USA

Penny is Associate Professor of Biological Sciences, Chair of the Education Committee of the Animal Behaviour Society, and Secretary of the International Society for Anthrozoology. Her training includes post-doctoral research at the Institute of Animal Behaviour, Rutgers University. She has specialized in field studies of animal social behaviour and communication in a variety of species, including prairie dogs, laughing gulls, humans, and cats. Penny is primarily interested in the role of communication in social groups, as revealed by following known individuals over time. She has undertaken additional laboratory work in the study of hormones and behaviour and ultrasonic communication. She teaches undergraduate major and non-major introductory biology courses in both cell and molecular biology and biological diversity, and also teaches animal behaviour.

John W. S. Bradshaw, BA (Oxon), PhD (Soton)
Director, Anthrozoology Institute, School of Clinical Veterinary Science, University of Bristol, Langford BS40 5DU, UK

John is Waltham Director of the Anthrozoology Institute, and Senior Lecturer in companion animal behaviour and welfare. His research interests include social, communication and olfactory behaviour of the domestic dog and cat, their welfare and its measurement; epidemiology and aetiology of behavioural disorders in companion animals; and characterisation of the pet-owner relationship. John is the author of "The Behaviour of the Domestic

Cat" (CAB International 1992). He was the cat specialist on the Council of Europe Carnivora Welfare Expert Group from 1998 to 2001 and a Cats Protection Council member from 1997 to 2003.

Rachel A. Casey, BVMS, Dip (AS) CABC, MRCVS
Deputy Director, Anthrozoology Institute, School of Clinical Veterinary Science, University of Bristol, Langford BS40 5DU, UK

Rachel is a veterinary surgeon with a Diploma in Companion Animal Behaviour Counselling. She is currently the Cats Protection Lecturer in Feline Behaviour and Welfare. Her research interests include the aetiology, epidemiology and treatment of clinical behaviour problems in the domestic cat, as well as the validation of methods to prevent the development of problem behaviours; the measurement of welfare parameters in the cat; measures of personality in the cat; and the validation of enrichment techniques for cats in rescue shelters.

Sharon L. Crowell-Davis, DVM, PhD, DACVB
Professor, Department of Anatomy and Radiology, College of Veterinary Medicine, University of Georgia, Athens, Georgia 30602, USA

Sharon is Professor of Veterinary Behaviour and Director of the Behaviour Service at the Veterinary Teaching Hospital of the University of Georgia, where she teaches all aspects of veterinary behaviour to veterinary students, clinical residents and graduate students. Her research interests include normal behaviour, and the causes and treatment of behaviour problems of cats, dogs, horses, parrots and rabbits. She has practiced clinical behaviour for 26 years and has published extensively in this field.

Sarah E. Heath, BVSc, MRCVS
Behavioural Referrals Veterinary Practice, 11 Cotebrook Drive, Upton, Chester CH2 1RA, UK

Sarah spent four years in veterinary practice before setting up a behaviour referral practice in 1992. She lectures at home and abroad on behavioural topics, and is an Honorary Lecturer at Liverpool University Veterinary School and a Recognised Teacher at Bristol University Veterinary School. She conducts behavioural clinics at the Bristol and Liverpool Schools and at private veterinary practices in the north-west of England. She is currently

Secretary for the British Small Animal Veterinary Association-affiliated Companion Animal Behaviour Therapy Study Group and President of the European Society for Veterinary Clinical Ethology. She is a contributing author and co-editor of the "BSAVA Manual of Canine and Feline Behavioural Medicine", and author of the "Henston Guide to Feline and Canine Behavioural Medicine". She has also written two popular books on feline behaviour, "Why does my cat?" and "Cat and Kitten Behaviour - an owner's guide".

Karyl J. Hurley, BSc, DVM, DACVIM, DECVIM-CA
Head of Academic Affairs, Waltham Centre for Pet Nutrition,1 Freeby Lane, Waltham-on-the-wolds, Melton Mowbray, Leicester LE14 4RT, UK

Karyl has a Bachelor of Science degree in Applied Physiology from Cornell University and a Doctorate in Veterinary Medicine from the New York State College of Veterinary Medicine. Her veterinary specialization included a small animal internship at Texas A&M University and a residency in small animal internal medicine at North Carolina State University. She was awarded a Diploma by the American College of Veterinary Internal Medicine in 1995 and the European College of Veterinary Internal Medicine in 1998. Karyl directs Global Scientific Communications at The Waltham Centre for Pet Nutrition since 1998. Recently, Karyl has undertaken a Visiting Fellow in Nutrition and Internal Medicine position at Cornell University whilst maintaining her role with Waltham.

Philip H. Kass, DVM, PhD, Diplomate ACVPM (Speciality in Epidemiology)
Associate Professor, Department of Population Health and Reproduction, School of Veterinary Medicine, University of California, Davis, California 95616, USA

Philip is Associate Professor of Epidemiology in the School of Veterinary Medicine and the School of Medicine at Davis. He teaches courses to graduate students in veterinary epidemiology, applied analytic epidemiology, advanced concepts in epidemiologic study design, and advanced topics in theoretical statistics. His research interests are companion animal epidemiology, biostatistics, and non-experimental inference. He is author or co-author of more than 175 publications, including a chapter on modern epidemiologic study designs in a forthcoming "Handbook of Epidemiology", to be published by Springer-Verlag.

Irene Rochlitz, BVSc, MSc, PhD, MRCVS
Animal Welfare and Human-animal Interactions Group, Department of Veterinary Medicine, University of Cambridge, Madingley Road, Cambridge CB3 0ES, UK

Irene is a research associate with the Animal Welfare and Human-animal Interactions Group at the University of Cambridge. She has a Master's degree in Veterinary Oncology and a PhD in Feline Welfare. She combines work in veterinary practice with research on issues affecting companion animals. Her interests include companion animal behaviour, the assessment of quality of life in companion animals, animal ethics and, in particular, the welfare of domestic cats.

Margaret R. Slater, DVM, PhD
Associate Professor, Department of Veterinary Anatomy and Public Health, College of Veterinary Medicine, Texas A&M University, College Station, TX 77843-4458, USA

Margaret is a companion animal epidemiologist with a long standing interest in pet overpopulation and companion animal welfare. Her research interests include free-roaming dog and cat issues and solutions. She is the author of the book "Community Approaches to Feral Cats", published by the Humane Society Press, and more than 70 scientific articles. She teaches epidemiology, biostatistics and evidence-based veterinary medicine to baccalaureate, veterinary and post-graduate levels.

Andreas Steiger, Prof., Dr. med. vet.
Professor of Animal Housing and Welfare, Division of Animal Housing and Welfare, Institute of Animal Genetics, Nutrition and Housing, Vetsuisse Faculty of the University of Bern, Bremgartenstrasse 109a, CH 3001 Bern, Switzerland

Andreas is a veterinary surgeon and head of the Division of Animal Housing and Welfare. He teaches animal housing and legislation on animal protection to students in Veterinary Medicine and in Zoology. His research interests include the housing, breeding, welfare and behaviour of companion animals and small pet mammals and birds, behavioural medicine and animal ethics. He is co-editor of a book on animal welfare (Sambraus H. and Steiger A., "Das Buch vom Tierschutz", 1997, Enke Verlag, Stuttgart).

Kit Sturgess, MA, VetMB, PhD, DSAM, CertVR, CertVC, MRCVS
Wey Referrals, 125-129 Chertsey Road, Woking, Surrey GU21 5BP, UK

Kit works in internal medicine in a referral veterinary practice. He holds a Royal Veterinary College Diploma in Small Animal Medicine, a Royal Veterinary College Certificate in Veterinary Radiology and a Royal Veterinary College Certificate in Veterinary Cardiology. His special interests are feline infectious disease and gastrointestinal disease, including nutrition and mucosal immunology. He is the author of the book "Notes on Feline Internal Medicine", published by Blackwell Science (2003).

Chapter 1

CAT BEHAVIOUR: SOCIAL ORGANIZATION, COMMUNICATION AND DEVELOPMENT

Sharon L. Crowell-Davis
Department of Anatomy and Radiology, College of Veterinary Medicine, University of Georgia, Athens, Georgia 30602, USA

Abstract: Whenever resources allow, feral cats form complex, matrilineal societies called colonies. Within the colony, there is extensive co-operation between adult females in the care and rearing of kittens, including communal nesting, grooming, and guarding. Males may either be closely associated with a given colony, or they may have home ranges that overlap with several colonies. Male and female cats of the same colony may have preferred conspecifics with which they engage in various affiliative behaviours. The sensitive period for social learning in kittens is between two and seven weeks of age. However, intraspecies social play peaks after this period, and examples of social learning have been observed in older kittens and juveniles. When caring for domestic cats, it is important that kittens are reared in a social environment that provides experiences with other kittens and adults so that they can learn appropriate, species-specific social behaviour.

1. INTRODUCTION

The ancestor of the domestic cat is the African wild cat, *Felis silvestris libyca*. It creates the core of a social group in the form of a queen and her kittens. In both the African wild cat and the domestic cat, *Felis silvestris catus*, kittens live with their mother for several weeks after birth until they are mature enough to have learned how to hunt on their own. If the kittens are raised in an environment where food resources are widely distributed or there are insufficient food resources to support many adults, the family group will disperse when the kittens are mature. Clumped food resources, consisting of food storage areas and refuse dumps, and rodent populations attracted to these resources, are a consequence of human civilization and provide the ecological context for the development of social organization in the domestic cat. These food sources contain enough food to support several small carnivores, and if these carnivores form a group they can defend their resource more successfully. When concentrated food sources are available, if

I. Rochlitz (ed.), The Welfare of Cats, 1–22.
© 2007 *Springer.*

the mother and kittens stay together, as the kittens become juveniles and the juveniles become adults, they are better able to defend the resource from strange cats. A group of related and familiar adults will provide formidable opposition to a lone cat attempting to access a concentrated resource such as the refuse from a fishing boat (Macdonald 1983; Frank 1988; Macdonald & Carr 1989). The formation of groups of related and familiar individuals around food sources is the first step in the development and organization of social behaviour in the domestic cat.

2. SOCIAL ORGANIZATION

A species is considered to be social "if their members live as enduring pairs, as families, or in larger groups in consequence of which social behaviour makes up a major proportion of their total activity" (Immelman & Beer 1989). During the past 25 years, many scientists in a number of countries and continents have studied groups of free-living or feral cats, called colonies, that maintain relatively consistent membership over long periods of time, exhibit individual recognition, engage in a variety of social interactions, and have a complex social organization (e.g. Dards 1978; Macdonald & Apps 1978; Dards 1983; Natoli 1985a, 1985b; Macdonald *et al.* 1987; Kerby & Macdonald 1988; Natoli & De Vito 1991; Mirmovitch 1995; Yamane *et al.* 1996; Sung 1998; Macdonald *et al.* 2000; Natoli *et al.* 2001; Wolfe 2001). Colony size is determined by existing food resources, and is directly proportional to food availability (Liberg *et al.* 2000).

Individual members of the colony recognize other members, as opposed to strangers. Integration of strangers is likely to be resisted, and there may be substantial disruption of the social order in the colony when a new cat joins the group. This occurs both in the free-living state, where a new cat chooses to stay rather than leave (e.g. Wolfe 2001) and in the domestic household, where a human imposes the integration on both the new cat and the existing colony.

Historically, cats were considered to be an asocial, solitary species that had no need for companionship and preferred to be alone (e.g. Milani 1987). In spite of the recent findings that free-living cats choose to live in groups, form close relationships with particular conspecifics ("preferred associates", see section 2.4.1) and engage in a variety of affiliative behaviours, this idea is still common in the popular press. This is unfortunate, since it often leads to the management and care of cats that is not in the best interests of their welfare. Inadequate understanding of the social needs and social relationships of cats can lead to problems such as excessive aggression and undesirable elimination (urination or defaecation) behaviour (see Chapter 4).

There is also the question of whether a cat's welfare is compromised when it is raised and maintained in a socially isolated situation, a not uncommon phenomenon. How much better off cats kept in stable family and social groups are, compared with cats kept in social isolation or in unstable social situations, remains to be addressed.

2.1 Colony Organization: Female Co-operation

The smallest colonies consist of a queen and her kittens that are not yet mature enough to survive independently. The heart of larger colonies is composed of several queens, often related, who help each other in a variety of ways to facilitate the survival of their young (Liberg & Sandell 1988; Macdonald *et al.* 2000). Queens assist each other during the process of parturition, a phenomenon not commonly observed among animals. The queen not giving birth will clean the perineum of the birthing queen, consume the fetal membranes and clean the newborn kitten (e.g. Macdonald & Apps 1978, Macdonald *et al.* 2000). Subsequently, queens groom, nurse and guard each other's kittens. Queens have also been observed to bring food to nursing queens in both *Felis silvestris catus* and the wild ancestor, *Felis silvestris libyca* (e.g. Smithers 1983; Macdonald *et al.* 1987). Nest-moving, when a queen moves her kittens from one nest to another, can be a particularly hazardous time for the kittens, as they may be left for extended periods. Kittens in communal nests spend less time alone during nest-moving, as at least one of the mothers is likely to be with them, than do kittens of queens raising a litter alone (Feldman 1993).

Reciprocal altruism is most likely to occur when the favours exchanged are of relatively equal value, and the exchange occurs fairly close in time (Trivers 1971). Queens living together in the same colony and giving birth to kittens in the same season, whether they are related or not, exist in ideal conditions for reciprocal altruism to be adaptive. If a queen who has given birth within the last few weeks or who is about to give birth within the next few weeks aids another queen in the care of her kittens, the effort she expends on the other queen's kittens will probably be reciprocated. In many cases the queens are related, for example they are mother and daughter, or sisters. In this case, they are also benefiting their own genetic fitness every time they engage in behaviours that increase the probability of survival of their siblings, grandchildren, or nieces and nephews (Hamilton 1964). The importance of the queen in maintaining the social group was shown in a study of a colony of neutered, adult cats in which the highest rates of allogrooming (see section 2.4.3) occurred in family groups where the mother was present. This occurred in spite of the fact that, at the time of the study, all family members were adults and had been neutered (Curtis *et al.* 2003).

2.2 Social Learning

Laboratory studies have revealed that cats are excellent observational learners. For example, they can learn arbitrary tasks that do not involve skills their ancestors needed to survive, simply by watching another cat complete the task (e.g. Chesler 1969). This ability is clearly relevant for kittens to rapidly learn hunting skills from their mother. The feral queen first brings dead prey to the nest, then live prey, beginning when the kittens are about four weeks old (Baerends-van Roon & Baerends 1979). After the queen has released live prey in the presence of her kittens, she will often demonstrate hunting techniques to them. The kittens will then gradually practise these tasks under the supervision of their mother. The relevance of the mother to learning in young kittens is further demonstrated by the fact that they socialize to humans more readily if their mother is present and calm during handling than if she is absent (Rodel 1986).

While most social learning undoubtedly is derived from the mother and surrogate mothers of a growing kitten and juvenile, males also appear to have a role. Adult males have been observed to use a forelimb to separate intensely wrestling juveniles, without engaging in aggression against them, to groom kittens, to curl up around kittens abandoned at a colony and to defend kittens (Macdonald 1987; Feldman 1993; Crowell-Davis et al. 1997, 2004).

While cats usually hunt game that is too small to be shared and hunt alone, the author has received reports from several researchers of observations of groups of two to four cats co-operatively hunting larger game such as rabbits or squirrels; this co-operative hunting included encirclement and subsequent sharing of the kill. Since only isolated instances, and not the development of this behaviour among groups of cats, have been observed, it is impossible to say if social learning as a juvenile is necessary for this behaviour to occur or if it can be learned as an adult.

Social learning in kittens, juveniles and adults is an area that has been, unfortunately, largely ignored. Adults teaching kittens and juveniles appropriate social skills, as well as appropriate hunting techniques, may be critical for the development of behaviourally functional colonies with individual members that are sociable and friendly, rather than aggressive.

2.3 Males: Alternative Strategies for Survival and Mating Success

Intense aggressive conflicts may occur between males, especially in the presence of an oestrus female, but even in this situation males do not necessarily fight (Figure 1). Within a colony, there are pairs of cats that are found close together less often than is typical for that colony. While no gender effect was found for this phenomenon in a colony of neutered cats, in a colony of intact cats such pairs were disproportionately male-male pairs, presumably because of sexual competition between them. However, there were also many intact male pairs that were "preferred associates", engaging in a range of affiliative behaviours (see 2.4.1) (Sung 1998; Wolfe 2001).

Figure 1. A male copulates with an oestrus female while another stands to the side. It was not known whether the two males were related, but their similar appearance suggested that they were. These cats and two other similar males remained near the female for an extended period without fighting. When one male had finished copulating, the female would move a few feet and then allow another male to mount. (Reprinted from J. Feline Medicine and Surgery 6, Crowell-Davis *et al.* Social organization in the cat: a modern understanding, 19-28. Copyright (2004), with permission from the European Society of Feline Medicine and the American Association of Feline Practitioners).

There is greater diversity in the life histories of males than of females. These histories can be grouped into two major categories. Some males spend most of their time with one particular group of females and develop strong social bonds with them. We might call these males "family males". In terms of lifetime reproductive success, there are several potential advantages to this strategy. First, females may be more willing to mate with familiar males than with unfamiliar males. Even if their overt mating activity may not show evidence of this, physiological mechanisms may exist that favour conception by the familiar male (Eberhard 1996). One study has identified that males who were members of a colony had greater mating success within that colony, even if they were small males, than did males from outside the colony, even if they were large (Yamane et al. 1996). Infanticide of kittens by males that are not members of the colony has been recorded (Pontier & Natoli 1999). Males who are members of the colony and the probable fathers of the kittens in that colony have been observed to assist the queens in aggressively driving off marauding males that present a danger to the kittens (e.g. Macdonald et al. 1987). Cats have a polygamous mating system: females mate with multiple males and males mate with multiple females. This causes an uncertainty of paternity, so that males should be unlikely to attack the kittens of any female they have mated with. An important aspect of male behaviour that is generally under-recognized is that males can, in fact, be quite friendly to kittens (Figure 2).

Figure 2. An intact male Bengal cat engages in wrestling play with his son. The adult male inhibits the intensity of his play so that the kitten is not harmed and, instead, learns appropriate species-specific play. (Courtesy of PrinceRoyal Bengals).

In contrast, other males do not form strong affiliations with particular groups or individual queens. Instead, they have large home ranges that overlap the home ranges of a number of different female groups. In this case, the strategy is probably to optimize the lifetime total mating success, although longitudinal studies of males adopting this "philandering" strategy as opposed to the "family" strategy have yet to be conducted. Larger males have an advantage when using the philandering strategy, as queens are least likely to mate with small, unfamiliar males (Yamane *et al.* 1996).

2.4 Affiliative Relationships and Behaviours

As mentioned previously, cats are not asocial; within a colony they may form close relationships with other cats. They show a variety of affiliative behaviours, including touching noses, allogrooming, allorubbing, playing and resting together.

2.4.1 Affiliative relationships

Within a colony, there are clearly identifiable preferred associates, particular cats that are found close to each other more often than they are to other conspecifics. Preferred associates do not simply gather at the same resource at the same time of day. Instead, they can be found together throughout the day, in a variety of contexts and locations (Wolfe 2001). Familiarity is important: as cats become more familiar with each other, they are more likely to remain close and allogroom (see section 2.4.3), and less likely to be aggressive (Barry & Crowell-Davis 1999; Curtis *et al.* 2003). Relatedness is even more important than familiarity, as related cats are more likely to be close and allogroom than unrelated cats of equal familiarity (Curtis *et al.* 2003).

2.4.2 Touching noses

Touching noses is a greeting behaviour, observed most commonly between preferred associates (Wolfe 2001, Figure 3). Males and females are equally likely to nose touch with the same or opposite gender (Sung 1998). Thus, individual, friendly relationships, rather than gender, determine the frequency of nose touches.

2.4.3 Allogrooming

In allogrooming one cat licks another, typically on the head or neck (Figure 4). It is more common between preferred associates than between non-preferred associates (Wolfe 2001). One cat may solicit allogrooming from another by approaching and lowering its head. Whether the groomer or the recipient of the grooming initiates the allogrooming, the cat being groomed is typically highly co-operative, steadily rotating the head and neck so that the groomer can reach various areas.

Figure 3. Two domestic cats touch noses in greeting. (Photo by Rebecca Knowles. Reprinted from J. Feline Medicine and Surgery 6, Crowell-Davis *et al.* Social organization in the cat: a modern understanding, 19-28. Copyright (2004), with permission from the European Society of Feline Medicine and the American Association of Feline Practitioners).

Figure 4. The cat in the middle lowers its head to solicit allogrooming from the cat on the right. (Photo by Rebecca Knowles. Reprinted from J. Feline Medicine and Surgery 6, Crowell-Davis *et al.* Social organization in the cat: a modern understanding, 19-28. Copyright (2004), with permission from the European Society of Feline Medicine and the American Association of Feline Practitioners).

2.4.4 Allorubbing

In allorubbing, two cats rub their heads, bodies and tails against each other (Figure 5). Allorubbing sessions may go on for several minutes. As with many behaviours, it probably has multiple functions. First, there is a tactile component. Allorubbing cats typically push and rub against each other quite vigorously, often purring as they do so. The social significance may be similar to the human hug, where the tactile contact serves to facilitate and maintain friendly social bonds. In addition, scent is probably exchanged during allorubbing. Allorubbing is more frequent among feral cats than among housecats that are kept together inside all of the time, and often occurs when feral cats are reuniting after separation to go hunting (e.g. Sung 1998; Barry & Crowell-Davis 1999).

2.4.5 Play

Play is a common and well-known behaviour of cats. It peaks between about four weeks to four months of age, but continues throughout adulthood. Wide individual variation occurs in cats, with some being extremely playful, playing with their owner, with other animals in the household, and on their own with any objects they can find to use as toys. On the other hand, some adult cats rarely play. A combination of genetics, types of experience, and timing of particular experiences with play probably determine this variation. To date, longitudinal studies to determine how the type and timing of early experience influences the degree of playfulness in adults have not been conducted.

It is notable that play occurs even between adult feral cats that live in poor nutritional conditions (Figure 6). Since play requires the expenditure of valuable calories, one would predict that, under such circumstances, play would not occur. The fact that it does raises the question of its relevance for social learning and the establishment and maintenance of social bonds.

Figure 5. Two feral cats engaging in side-to-side allorubbing. (Photo by Rebecca Knowles. Reprinted from J. Feline Medicine and Surgery 6, Crowell-Davis *et al.* Social organization in the cat: a modern understanding, 19-28. Copyright (2004), with permission from the European Society of Feline Medicine and the American Association of Feline Practitioners).

Figure 6. Two feral cats engaging in play behaviour. These adults are small due to chronic malnutrition, yet still expend energy on social play. (Photo by Rebecca Knowles. Reprinted from J. Feline Medicine and Surgery 6, Crowell-Davis *et al.* Social organization in the cat: a modern understanding, 19-28. Copyright (2004), with permission from the European Society of Feline Medicine and the American Feline Association of Practitioners).

2.4.6 Resting Together

Cats do not only engage in active social behaviours such as those discussed previously. Both in feral colonies and in household groups, cats can often be found resting together in close physical contact (Figure 7). This behaviour occurs even when there is sufficient space for the cats to spread out. It does not have a thermoregulatory function, as it is seen in feral colonies on hot, humid days. Cats choose to rest together because of strong, specific social bonds.

Figure 7. Unrelated intact male and female cats rest together. The female was not in oestrous, other rooms and soft surfaces were available, and the room was not cold. (Courtesy of PrinceRoyal Bengals).

2.5 Social Conflict, Dominance Relationships and Control of Resources

While there are benefits to group living, there are also disadvantages. Group living provides easy access to potential mates, but also provides access to sexual competitors. The group may exist because it can effectively defend a food resource against outsiders, but as the group's size becomes too large, there is competition for food. Even in a domestic household where the human caregivers increase the amount of food available to allow for increasing numbers of cats, there may be competition for the food resource at particular times and locations. Elimination sites are probably not typically a source of competition in the feral state, but in the domestic household where only one or a few litter boxes may be offered, access to desirable elimination sites can become a source of fierce competition.

The formation of dominance hierarchies or, in many ways a more useful term, subordinance hierarchies can mediate access to resources and the expression of aggression. When an individual typically defers, or gives way, to another individual, and when that deferral is a consequence of past interactions between those two individuals, the animal that defers is said to be subordinate, while the animal that is deferred to is said to be dominant (e.g. Bernstein 1981; Immelman & Beer 1989). When we know the dominance relationships between multiple pairs within a group, we can construct a dominance hierarchy.

Throughout the animal kingdom, truly linear hierarchies are extremely rare, except in small groups of animals numbering five or less (Lehner 1996). Instead, ties and reversals are common, especially in larger groups. While strictly linear hierarchies may be found in small groups of cats, the typical large colony is likely to have a complex hierarchy that is only partially linear (e.g. Natoli & de Vito 1991; Knowles 2003).

Cats signal deference or submission in a variety of ways, including walking around another cat, waiting for another cat to pass before moving into an area, retreating when another cat approaches, and avoiding eye contact. Body, tail and ear postures include hunching, crouching, rolling over on the back, tucking the tail to either side of the thigh and turning the ears down or back (Feldman 1994b; Knowles 2003). Dominant cats will block the movement of subordinate cats or even supplant them. They will feint or bat at the subordinate with their paw, chase and sometimes mount the subordinate, although this behaviour does not appear to be as common in cats as it is in dogs. They will also signal with body posture, including holding the ears up and rotated to the side, arching the base of the tail and staring at the subordinate (Figure 8). In an encounter, the cat may exhibit a complete dominance display, with the hind limbs extended and stiff, the base of the tail elevated with the remainder hanging, and the ears stiffly erect and rotated laterally while it stares at the subordinate. In extreme displays, the dominant cat will slowly wag its head from side to side. In a group of cats that get along well, however, most dominance displays are more subtle and involve only one or two components of the complete display. For example, if a dominant cat is walking past a subordinate it may briefly stare at it while rotating its stiffly erect ears to the side. If the subordinate cat averts its gaze, the dominant cat will then walk on.

Figure 8. Two cats engage in a "stare-down" which resulted in the cat on the right deferring, by breaking eye contact and moving away. (Photo by Rebecca Knowles. Reprinted from J. Feline Medicine and Surgery 6, Crowell-Davis *et al.* Social organization in the cat: a modern understanding, 19-28. Copyright (2004), with permission from the European Society of Feline Medicine and the American Association of Feline Practitioners).

It is important to understand dominance relationships between cats, because they can have profound effects on the rate of aggression and access to preferred resources. Cats that are higher-ranking in social contexts away from food can control access to food by subordinates, although they do not always do so (Knowles 2003) (see Chapters 2 and 4 for other views on dominance and hierarchies in cats).

How hungry the higher-ranking and lower-ranking cats are, as well as absolute rank, no doubt affect the outcome of a particular interaction at a food source. Likewise, access to the litter box may be controlled by either the highest-ranking cat in the household or, collectively, by a group of high-ranking cats. The author encounters this phenomenon commonly in her clinical practice. Control of important resources by high-ranking cats obviously has important implications for the welfare of subordinate cats. In groups of cats where there is social harmony, the high-ranking cats typically are not aggressive, so long as subordinate cats defer whenever a more dominant cat attempts to access a space or resource. In other groups, one or more high-ranking cats may habitually and frequently threaten lower-ranking cats, even when they do not appear to want a particular resource.

They may also monopolize a resource for substantial lengths of time, even if they are not using it. For example, in one of the author's clinical cases, the dominant cat in the household spent much of the day sitting and lying in the doorway to the room where the litter boxes were, and would not allow the subordinate cat to pass. This subordinate cat therefore eliminated in an area of the house far away from the litter boxes. Adding a litter box to this area resolved the problem. Persistent aggression by high-ranking cats against subordinates, or confiscation of resources by high-ranking cats, produces chronic social tension and conflict (see Chapters 2 and 4).

The experiential and genetic factors that produce one type of cat or another are not well understood, and require further study if we are to identify how to best keep cats in groups where social harmony is the norm. Inadequate and/or inappropriate early socialization experiences no doubt contribute to the production of the 'bully' or 'despot' cat. Crowding, such as often occurs in the laboratory colonies where cat behaviour has often been studied, is probably a contributing factor to such behaviour. In the colony studied by Knowles (2003), cats frequently exhibited ritualized dominance and submission, but genuine aggression was extremely rare. This is what would be expected in social groups where the hierarchy is stable. The only real fight ever observed occurred between two neutered males who were close in age and weight, and whose dominance relationship was unresolved, as they had approximately equal 'wins' and 'losses' against each other. Clearly, failure to resolve who is dominant can be a source of conflict.

In spite of the frequent references to territoriality in the cat, there is no good evidence that cats are territorial. Animals that are territorial do not just have a home range, but actively and consistently defend their specific piece of land. Their home range is synonymous with their territory, and they leave prominent signals on its borders. Adjacent home ranges do not over-lap (Bernstein 1981; Lehner 1996). The home ranges of cats commonly overlap, and Feldman (1994a) found that the deposition of urine and faeces, and the creation of scratching sites, occurred well within the home range along the cat's most commonly used travelling routes, rather than on the borders.

3. COMMUNICATION

3.1 Visual Communication

Cats communicate mood and intention by a variety of visual signals. A visual signal that indicates friendly intentions is the 'tail-up' position, in which the cat approaches another cat, human, dog or other animal with the tail sticking up in the air, perpendicular to the ground. Allorubbing occurs most frequently if one or both cats that approach each other do so in the tail-up position (Cameron-Beaumont 1997). Cats will also give a variety of other social signals visually, including solicitation of grooming, and ritualized dominance and submission signaling, as discussed previously.

At any given moment, a cat is giving visual signals through body position and movement. Cats also create long-term visual signals that remain even after the cats have gone. The most common and prominent of these is generated by scratching on surfaces, including vertical and horizontal tree trunks, posts, logs and similar objects (see Chapter 4). The scratched area leaves the information that the individual that made the mark was strong enough and healthy enough to have done the damage. As mentioned previously, scratched areas are usually generated along the routes most commonly used within a cat's home range, rather than along the periphery (Feldman 1994a). They may serve as landmarks that the cat uses in navigating through its home range.

3.2   Tactile Communication

Cats that are members of the same colony are often in tactile contact through such behaviours as allorubbing, nose touching and resting together. Using each other as 'pillows' is a common phenomenon among both feral and house cats. The tail is sometimes actively used to stroke another cat, and cats may back up to each other and wrap their tails together (Crowell-Davis 2002). The exact meaning of this behaviour is not well understood, although it appears to be involved in social bonding. The frequency and intensity of physical contact is probably a good measure of social bonding between cats, and may be useful for future studies that focus on identifying management techniques that maximize social harmony within a group.

3.3   Vocal Communication

Cats have three major types of vocal communication. Within each type are numerous variations, and the total number of specific vocalizations is substantial, although the exact number will vary depending on whether one tends to 'split' or 'lump' various subcategories. Regardless, cats have one of the most extensive repertoires of any carnivore species, probably because they must often communicate with conspecifics in dim light or where visibility is poor for other reasons, such as in a brushy area (Moelk 1944; Kiley-Worthington 1984).

The purr and the trill are formed with the mouth closed. The trill is a greeting call, while the purr occurs in a wide variety of forms and contexts as a friendly greeting and care-soliciting call. It will occur between cats during a variety of amicable social interactions, including allogrooming, allorubbing and resting together.

The miaow is formed with the mouth initially open and then gradually closing. The miaow is perhaps the most highly varied call, with many different forms. It occurs in a variety of situations during friendly interactions with other cats. Humans are very responsive to the miaow and so, by reinforcement, it often becomes the signal that a cat uses to indicate that it wants to be fed, petted, let out of the house, or some other care-giving action of a human.

The growl, yowl, snarl, hiss, spit and shriek are sounds that are made with the mouth held open in a relatively constant position. These sounds are all related to aggression of various sorts, including intraspecies conflict and defensive aggression against threatening predators or humans.

3.4 Olfactory Communication

Cats have an excellent sense of smell, and the importance and function of olfactory communication is doubtless inadequately understood because of the poor sense of smell of human researchers. Sebaceous glands are located throughout the body, especially on the head, in the perianal area, and between the digits. The frequency with which cats rub and sniff each other supports the idea that olfactory cues from these areas are important. Urine and faeces also appear to be used in olfactory communication, as does scratching (scent is deposited from the inter-digital glands).

Cats frequently rub their heads on each other, often purring as they do so, and on objects in their home range. Some pet cats that have close relationships with their owner will rub their perioral region back and forth across the perioral region of the owner, also purring as they do when engaging in such interactions with conspecifics. The cat's behaviour suggests that it is depositing scent on conspecifics with which it has a friendly relationship, and depositing scent within the home range.

Urine spraying is a poorly understood behaviour in which the cat backs up to a vertical object, raises its tail, and sends a typically small quantity of urine backwards onto the object. It probably has multiple meanings, depending upon context and the exact molecules within a given spray of urine. Urine spraying is commonly attributed to being a form of territorial marking. However, as discussed previously, cats are not territorial animals. Based on various field reports on the deposition of urine and cats' responses to it, urine spraying probably gives identifying information, particularly about reproductive status, about which cat was at a given location at a given time, as well as about emotional state, for example if the sprayer was aroused (Verberne & de Boer 1976; Natoli 1985b; Gorman & Trowbridge 1989; Passanisi & Macdonald 1990; Feldman 1994a) (see Chapter 4).

Feral cats have been observed to bury faeces in the core, or in the most commonly used, area of their home range, but occasionally leave faeces unburied in both the core areas (personal observation) and the periphery of their home range (e.g. Macdonald *et al.* 1987; Feldman 1994a). Sung (2001), in a study of domestic cats, did not find any relationship between elimination behaviour problems, including leaving faeces exposed, and the location of the litter box in the home, that is whether it was in the cats' core area of activity or in a peripheral area.

4. DEVELOPMENT OF BEHAVIOUR IN KITTENS

Cats are altricial, kittens being born with closed eyes and ears, and poorly developed auditory and visual systems (Gottlieb 1971). They are unable to urinate or defecate without the assistance of their mother and are totally dependent on her for nourishment. For the first two weeks of life, they can only move with a slow, paddling gait. Opening of the eyes typically occurs at seven to 10 days, although it can happen as early as two days and as late as 16 days in normal kittens (Villablanca & Olmstead 1979). A variety of genetic and environmental factors affect the timing of eye opening, including gender, paternity, exposure to light, and age of the mother at the time of parturition (Braastad & Heggelund 1984). Kittens typically begin to walk during the third week of life, and run during the fifth week (Moelk 1979).

Maturation is a gradual process, affected by a variety of genetic, nutritional and environmental factors. The feral queen will begin bringing prey to her offspring at about the fourth week of life, which suggests that this is when kittens are first capable of learning complex tasks (Baerends-van Roon & Baerends 1979). Handling kittens produces decreased fearfulness and faster development of a variety of physical and behavioural traits (e.g. Meier 1961; Wilson *et al.* 1965). Behavioural and physical development is delayed if the mother is malnourished (e.g. Smith & Jansen 1977a, b; Simonson 1979; Gallo *et al.* 1980, 1984). Some of these effects are directly due to the malnourishment, as certain areas of the kitten's brain are underdeveloped. However, some effects are probably due to learning deficiencies, as queens who have been undernourished during pregnancy and lactation are more aggressive to their kittens than well-fed queens. As these deleterious effects continue into the second generation (Simonson 1979), this phenomenon should be borne in mind when attempting to domesticate the kittens of feral queens that have been chronically malnourished.

Play is an important component of the early development of kittens. Social play, which may be practiced for later social interactions, peaks in the third and fourth month of life. Object play, which probably functions primarily to facilitate the development of skills that will be useful in hunting, peaks around the fifth month of life (Mendoza & Ramirez 1987). Singleton kittens have less experience with social play than do kittens with siblings, even though their mothers do engage in some play with them (Mendl 1988).

5. SOCIALIZATION

As with all social species, the cat is born with the ability to learn social skills, given the appropriate social environment. However, it is not born with specific social skills, and the kitten that is raised as a singleton from birth to five or six weeks of age, then taken away from its mother and raised as an only cat, is likely to be profoundly socially incompetent. It will be an asocial individual, not because this is normal for the species but because of how it was raised.

The sensitive period for socialization is the period of time during which an animal is particularly susceptible to social learning experiences (Immelmann & Beer 1989). Little research has been conducted on sensitive periods and specific social learning in the cat, probably because of the misconception that they are not social. Karsh (1983) found that the best socialization to humans occurs if kittens are handled from the second to the seventh week of life. The presence of a calm mother is beneficial to the process (Rodel 1986) (see Chapters 3 and 4). In order to facilitate the formation of socially stable groups in cats, research needs to be conducted on when socialization to their own species occurs, and what particular early experiences are necessary for the development of high rates of affiliative behaviour and social attachment, and low rates of aggression. Until such research is completed, it is reasonable to assume that the period of socialization of kittens to other cats also occurs between 2 to 7 weeks of age.

6. BEHAVIOUR DIRECTED TOWARDS HUMANS

Intraspecies social behaviour forms the foundation for social behaviour directed at humans. It is specifically because they are a social species that cats have evolved from being creatures that simply live off the pests in human refuse, such as rodents, to being companion animals that share our homes and beds. Much of the social behaviour directed at humans echoes the social behaviour that occurs between cats that are members of the same colony. As noted previously, allorubbing is most common among feral cats, that are often not together continuously, and occurs particularly when a cat returns from a hunt (purring being a common vocalization during allorubbing). Rubbing on humans' legs is transference of this behaviour, which establishes social bonding between the cat and another species. Thus, when a human returns home from work and their cat rubs their legs, the cat is engaging in a species-typical friendly greeting that it would also engage in with a conspecific returning to the heart of the colony. Cats that are friendly to dogs can be observed to rub on them. The propensity of cats that are well socialized to humans to lie in their lap or against their thigh is transference of another species-typical social behaviour to an interspecies situation. Similarly, cats approaching humans in a friendly fashion are likely to have their tails sticking straight up in the 'tail-up' position, as discussed in section 3.1, and may wrap their tail around a human's calf in a fashion reminiscent of intraspecies tail-wrapping. Purring, a care-soliciting and greeting vocalization that commonly occurs between cats that are friendly with each other, is fostered by humans who give attention when the cat purrs. Many humans enjoy these behaviours, which then form the basis for humans becoming attached to cats, caring for them, and keeping them in their homes (see Chapter 3).

7. CONCLUSIONS

The domestic cat is a species that, given the right conditions, forms complex, matrilineal social groups. Relatedness, familiarity and individual histories are all relevant to issues of social bonding, affiliative behaviour, stability or disruption of the hierarchy, and the incidence of aggression. Learning of social behaviour occurs between 2 and 7 weeks of age (the sensitive period for socialization to humans) and beyond, as within-species social play peaks several weeks later. Social behaviours that are exhibited within the species are redirected to humans and other animals, and form the basis for interspecies social bonding. While our understanding of the social behaviour and social needs of cats has increased substantially in the past two decades, many questions remain inadequately answered. If we are going to keep cats as companion animals, it is imperative that we develop a better understanding of their social needs so that we can care for them in such a way as to maximize their welfare.

8. REFERENCES

Baerends–van Roon, J.M. and Baerends, G.P. (1979) The genesis of the behaviour of the domestic cat, with emphasis on the development of prey catching. *Verhaelingen der Koninklijke Nederlandse Akademie van Wetenschappen Afd. Natuurkunde, Tweede Reeks, Deel* **72**, 1-115.

Barry, K.J. and Crowell-Davis, S.L. (1999) Gender differences in the social behavior of the neutered indoor-only domestic cat. *Applied Animal Behaviour Science* **64**, 193-211.

Bernstein, I.W. (1981) Dominance: the baby and the bathwater. *The Behavioral and Brain Sciences* **4**, 419-457.

Braastad, B.O. and Heggelund, P. (1984) Eye-opening in kittens: effects of light and some biological factors. *Developmental Psychobiology* **17**, 675-681.

Cameron Beaumont, C.L. (1997) Visual and tactile communication in the domestic cat (*Felis silvestris catus*) and undomesticated small felids. PhD thesis, University of Southampton.

Chesler, P. (1969) Maternal influence in learning by observation in kittens. *Science* **166**, 901-903.

Crowell-Davis, S.L. (2002) Social behaviour, communication and development of behaviour in cats. In Horwitz, D.F., Mills, D.S. and Heath, S. (eds.). *BSAVA Manual of Canine and Feline Behavioural Medicine*, British Small Animal Veterinary Association, Quedgeley, Gloucester, pp. 21-29.

Crowell-Davis, S.L., Barry, K. and Wolfe, R. (1997) Social behavior and aggressive problems of cats. *Veterinary Clinics of North America: Small Animal Practice* **27**, 549-568.

Crowell-Davis, S.L., Curtis, T.M. and Knowles, R.J. (2004) Social organization in the cat: a modern understanding. *J. Feline Medicine and Surgery* **6**, 19-28.

Curtis, T.M., Knowles, R.J. and Crowell-Davis, S.L. (2003). Influence of familiarity and relatedness on proximity and allogrooming in domestic cats (*Felis catus*). *American J. Veterinary Research* **64**, 1151-1154.

Dards, J.L. (1978). Home ranges of feral cats in Portsmouth Dockyard. *Carnivore Genetics News Letter* **3**, 242-255.

Dards, J.L. (1983) The behaviour of dockyard cats: Interactions of adult males. *Applied Animal Ethology*, **10**, 133-153.

Eberhard, W.G. (1996) *Female Control: Sexual selection by Cryptic Female Choice.* Princeton University Press, Princeton, New Jersey.

Feldman, H.N. (1993) Maternal care and differences in the use of nests in the domestic cat. *Animal Behaviour* **45**, 13-23.

Feldman, H.N. (1994a) Methods of scent marking in the domestic cat. *Canadian J. Zoology* **72**, 1093-1099.

Feldman, H.N. (1994b) Domestic cats and passive submission. *Animal Behaviour* **47**, 457-459.

Frank, S.A. (1998) *Foundations of Social Evolution.* Princeton University Press, Princeton, New Jersey.

Gallo, P.V., Werboff, J. and Knox, R. (1980) Protein restriction during gestation and lactation: development of attachment behaviour in cats. *Behavioural and Neural Biology* **29**, 216-223.

Gallo, P.V., Werboff, J. and Knox, R. (1984) Development of home orientation in protein-restricted cats. *Developmental Psychobiology* **17**, 437-449.

Gorman, M.L. and Trowbridge, B.J. (1989) The role of odor in the social lives of carnivores. In Gittelman, J.L. (ed.). *Carnivore Behavior, Ecology and Evolution*, Chapman and Hall, London, pp. 55-88.

Gottlieb, G. (1971) Ontogenesis of sensory function in birds and mammals. In Tobach, E., Aronson, L.R. and Shaw, E. (eds.). *The Biopsychology of Development*, Academic Press, New York, pp. 66-128.

Hamilton, W.D. (1964) The genetical evolution of social behaviour. II. *J. Theoretical Biology* **7**, 17-52.

Immelman, K. and Beer, D. (1989) *A Dictionary of Ethology*. Harvard University Press, Cambridge, Massachusetts, p. 273.

Karsh, E.B. (1983) The effects of early handling on the development of social bonds between cats and people. In Katcher, A.H. and Beck, A.M. (eds.). *New Perspectives on our Lives with Companion Animals*, University of Pennsylvania Press, Philadelphia, pp. 22-28.

Kerby, G. and Macdonald, D.W. (1988). Cat society and the consequences of colony size. In Turner, D.C. and Bateson, P. (eds.). *The Domestic Cat: the biology of its behaviour*, 1st edn., Cambridge University Press, Cambridge, pp. 67-82.

Kiley-Worthington, M. (1984) Animal language? Vocal communication of some ungulates, canids and felids. *Acta Zoologica Fennica* **171**, 83-88.

Knowles, R. (2003) Correlation of dominance based on agonistic interactions with feeding order in the domestic cat (*Felis catus*). M.Sc. thesis. University of Georgia, Athens, GA.

Lehner, P.N. (1996) *Handbook of ethological methods*, 2nd edn., Cambridge University Press, Cambridge.

Liberg, O. and Sandell, M. (1988) Spatial organization and reproductive tactics in the domestic cat and other felids. In Turner, D.C. and Bateson, P. (eds.). *The Domestic Cat: the biology of its behaviour*, 1st edn., Cambridge University Press, Cambridge, pp. 83-98.

Liberg, O., Sandell, M., Pontier, D. and Natoli, E. (2000) Density, spatial organization and reproductive tactics in the domestic cat and other felids. In Turner, D.C. and Bateson, P. (eds.). *The Domestic Cat: the biology of its behaviour*, 2nd edn., Cambridge University Press, Cambridge, pp. 119-147.

Macdonald, D.W. (1983). The ecology of carnivore social behaviour. *Nature* **379**, 379-384.

Macdonald, D.W. and Apps, P.J. (1978). The social behaviour of a group of semi-dependent farm cats, *Felis catus*: A progress report. *Carnivore Genetic News Letter* **3**, 256-268.

Macdonald, D.W., Apps, P.J., Carr, G.M. and Kirby, G. (1987) Social dynamics, nursing coalitions and infanticide among farm cats, *Felis catus*. *Advances in Ethology* (supplement to *Ethology*) **28**, 1-66.

Macdonald, D.W. and Carr, G.M. (1989). Food security and the rewards of tolerance. In Standen, V. and Folley, R.A. (eds.). *Comparative socioecology: the behavioural ecology of human and other mammals*, Blackwell Scientific Publications, Oxford, pp. 75-99.

Macdonald D.W., Yamaguchi, N. and Kerby, G. (2000). Group-living in the domestic cat: its sociobiology and epidemiology. In Turner, D.C. and Bateson, P. (eds.). *The Domestic Cat: the biology of its behaviour*, 2nd edn., Cambridge University Press, Cambridge, pp. 95-118.

Meier, G.W. (1961) Infantile handling and development in Siamese kittens. *J. Comparative Physiology and Psychology* **54**, 284-286.

Mendoza, D.L. and Ramirez, J.M. (1987) Play in kittens (*Felis domesticus*) and its association with cohesion and aggression. *Bulletin of the Psychonomic Society* **25** (1), 27-30.

Mendl, M. (1988) The effects of litter-size variation on the development of play behaviour in the domestic cat: litters of one and two. *Animal Behaviour* **36**, 20-34.

Milani, M. (1987) *The Body Language and Emotion of Cats*. William Morrow and Company, Inc., New York.

Mirmovitch, V. (1995) Spatial organization of urban feral cats (*Felis catus*) in Jerusalem. *Wildlife Research* **22**, 299-310.

Moelk, M. (1944) Vocalizing in the house-cat: A phonetic and functional study. *American J. Psychology* **57**, 184-205.

Moelk, M. (1979) The development of friendly behavior in the cat: a study of kitten-mother relations and the cognitive development of the kitten from birth to eight weeks. *Advances in the Study of Behavior* **10**, 164-224.

Natoli, E. (1985a) Spacing pattern in a colony of urban stray cats (*Felis catus* L.) in the historic centre of Rome. *Applied Animal Behavior Science* **14**, 289-304.

Natoli, E. (1985b) Behavioural responses of urban feral cats to different types of urine marks. *Behaviour* **94**, 234-243.

Natoli, E., Baggio, B. and Pontier, D. (2001) Male and female agonistic and affiliative relationships in a social group of farm cats (*Felis catus* L.). *Behavioural Processes* **53**, 137-143.

Natoli, E. and de Vito, E. (1991) Agonistic behaviour, dominance rank and copulatory success in a large multi-male feral cat, *Felis catus* L., colony in central Rome. *Animal Behaviour* **42**, 227-241.

Passanisi, W.C. and Macdonald, D.W. (1990) Group discrimination on the basis of urine in a farm cat colony. In Macdonald, D.W., Muller-Schwarze, D. and Natynczuk, W.E. (eds.). *Chemical Signals in Vertebrates* 5, Oxford University Press, Oxford, pp. 337-345.

Pontier, D. and Natoli, E. (1999) Infanticide in rural male cats (Felis catus L.) as a reproductive mating tactic. *Aggressive Behavior* **25**, 445-449.

Rodel, H. (1986) Faktoren, die den Aufbau einer Mensch-Katze-Beziehung beeinflussen. PhD Thesis, University of Zurich-Irchel, Switzerland.

Simonson, M. (1979) Effects of maternal malnourishment, development and behavior in successive generations in the rat and cat. In Levitsky, D.A. (ed.). *Malnutrition, Environment and Behavior*, Cornell University Press, Ithaca, New York, pp. 133-160.

Smith, B.A. and Jansen, G.R. (1977a) Brain development in the feline. *Nutrition Reports International* **16**, 487-495.

Smith, B.A. and Jansen, G.R. (1977b) Maternal undernutrition in the feline: brain composition of offspring. *Nutrition Reports International* **16**, 497-512.

Smithers, D.H.N. (1983). *The Mammals of the Southern African Subregion*. University of Pretoria, Pretoria, South Africa, p. 390.

Sung, W. (1998) Effect of gender on initiation of proximity in free ranging domestic cats (*Felis catus*). M.Sc. Thesis, University of Georgia, Athens, GA.

Sung, W. (2001) The elimination behavior patterns of domestic cats (*Felis catus*) with and without elimination behavior problems. Ph.D. Dissertation, University of Georgia, Athens, GA.

Trivers, R.L. (1971) The evolution of reciprocal altruism. *Quarterly Review of Biology* **46**, 35-57.

Verberne, G. and de Boer, J. (1976) Chemocommunication among domestic cats, mediated by the olfactory and vomeronasal senses. *Zeitschrift fur Tierpsychologie* **42**, 86-109.

Villablanca, J.R. and Olmstead, C.E. (1979) Neurological development in kittens. *Developmental Psychobiology* **12**, 101-127.

Wilson, M., Warren, J.M. and Abbot, L. (1965) Infantile stimulation, activity, and learning in cats. *Child Development* **36**, 843-854.

Wolfe, R. (2001) The social organization of the free ranging domestic cat (*Felis catus*). Ph.D. Dissertation, University of Georgia, Athens, GA.

Yamane, A., Doi, T. and Ono, Y. (1996) Mating behaviors, courtship rank and mating success of male feral cats *(Felis catus)*. *J. Ethology* **14**, 35-44.

Chapter 2

THE ASSESSMENT OF WELFARE

Rachel A. Casey and John W. S. Bradshaw
Anthrozoology Institute, School of Clinical Veterinary Science, University of Bristol, Langford BS40 5DU, UK

Abstract: This chapter adopts the perspective that welfare is based upon subjective states experienced by the animal, i.e. those equated with emotions and feelings. While current methods do not allow us to access any animal's mental state directly, indirect approaches can be used. One approach attempts to assess the welfare of a cat using behavioural and physiological measures of stress, while another predicts welfare from the degree of 'fit' between the cat's current environment and the environment in which its ancestral species evolved. Behavioural measures have not been entirely consistent from one study to another, but a composite scale, the Cat-Stress-Score, has been developed for cats kept in small enclosures, and approach tests to assess the impact of specific stressors have also been used. Of the possible physiological indicators of stress, urinary cortisol has been measured most often. Better integration between behavioural and physiological indicators, alternative measures to account for individual differences, and ways of addressing cognitive processes directly, may all offer ways of furthering the scientific assessment of welfare. Furthermore, in practice the procedures adopted for welfare assessment of cats often differ, depending upon whether the subject is presented by its owner for a behavioural problem, or is kept in confinement in a shelter, cattery or laboratory, and integration of these procedures should lead to further improvements.

1. INTRODUCTION

1.1 Definition of Animal Welfare

Animal welfare, in its broadest sense, can be considered from three interacting but distinct perspectives (Duncan & Fraser 1997). First, welfare may be defined purely in terms of biological functioning, for example the animal's health, absence of injury, and capacity to reproduce. This is the viewpoint historically adopted by veterinary scientists and will not be considered in detail in this chapter. The second perspective assumes that the animal's welfare is determined by whether, and how closely, it is able to perform behaviours that are 'natural'. These are not always easy to define for a domesticated species, but this perspective can lead to valuable insights. The third perspective, which we will predominantly use in this chapter, is based on the premise that animals have feelings and emotions, such as pleasure, fear and anxiety, that are in some way analogous to our own. It is these feelings and emotions, a reflection of the animal's mental state, that determine the animal's welfare. Because welfare arises from intrinsically 'invisible' processes that are ultimately private to the animal that experiences them, there are considerable practical difficulties in assessing it. While in recent years considerable progress has been made in devising methods for probing the mental states of non-human animals (e.g. Mason *et al.* 2001; Harding *et al.* 2004), they have not been tested in cats.

There are two main approaches to establishing the welfare status of an animal; it is assumed that both give us an understanding of the animal's feelings and emotions, that is its mental state. The approach most commonly used looks at various parameters related to the individual animal, such as its behaviour and physiological state, and uses these to assess to what extent the individual is coping with the environment in which it finds itself (Broom 1988). The other approach, which can be regarded as arising from the 'natural behaviour' perspective mentioned previously, looks at the species level in order to ascertain the environmental features that are likely to influence welfare, because each species is uniquely adapted to thrive in different environmental circumstances through natural and artificial selection (Barnard & Hurst 1996). By comparing the environment in which the species has adapted to survive with that which it currently occupies, we can make some predictions about its likely level of welfare.

Both of these approaches have advantages and disadvantages, and it is beyond the scope of this chapter to discuss these in detail. It is relevant, nevertheless, to consider that in the first approach individual differences in both behavioural strategy and physiological parameters can make the assessment of single animals difficult (Mason & Mendl 1993), and in the second the identification of a 'natural' environment is not straightforward when considering a domesticated species.

In practice, it is sensible to assess welfare with both of these approaches in mind: each animal should be assessed in terms of comparing its environment with an optimal environment for its ethological and motivational needs, as well as through measurement of individual coping strategies and the degree to which they are successful. In this chapter, we will look at the different approaches to measuring welfare in the domestic cat and identify the practical and interpretative advantages and problems for each.

Much of the research on cat welfare has been conducted on animals in confinement, mainly those housed in shelters, catteries and in laboratory colonies. These settings appear to have been chosen partly for convenience, because they contain populations kept under similar conditions. While cat welfare under these circumstances is undoubtedly important, and ensuring good welfare should form part of the duty of care for those running such establishments, it must not be forgotten that there are also welfare concerns for cats living under other circumstances. Very little attention has been paid to the welfare of strays and cats in feral colonies (Figure 1), even though such cats may be 'rescued' on the assumption that their welfare can be improved. The welfare of pet cats has, so far, largely been approached from a clinical perspective, by the assessment of the mental states of cats presented to clinicians by their owners for problematic behaviour. The integration of this approach with that used for cats in confinement has scarcely begun, but may prove valuable.

Figure 1. A litter of feral unsocialized kittens during "rescue". The welfare implications of bringing such animals into close contact with humans have yet to be fully evaluated. (Courtesy of Sarah Lowe).

1.2 The Stress Response

The stress response is an adaptive mechanism that enables an animal to react rapidly to an event that changes its homeostatic status. In everyday use, the term 'stress' is used to refer both to the physiological response described by Selye (1956) and, particularly within the human context, to an event or situation that causes a chronic negative impact on behaviour, health and welfare. In scientific terms, the 'stress response' is strictly the physiological response that occurs to a range of emotional and motivational changes, and those events or situations that precipitate an acute or chronic stress response are termed stressors.

Although we are discussing the stress response in terms of its negative impact on welfare, it is important to remember that the stress response is a normal and highly adaptive mechanism in cats. It initiates changes that provide the individual with the resources for immediate skeletal activity, including optimizing vigilance and responsiveness (Weipkema & Koolhaas 1992) so that it can behave appropriately towards a change in the external environment. It also allows resources for an internal event, such as the response of the immune system to an internal challenge (Ader & Cohen 1993). During the stress response, the autonomic output increases sympathetic activity and decreases parasympathetic activity (Schwaber et al. 1982). The sympathetic autonomic activity initiates all those changes that are familiarly associated with the 'fight or flight' response, such as increases in heart rate, cardiac output, respiratory rate and vasodilation to vital organs, and decreases in gastrointestinal and reproductive organ activity, which in combination prepare the body for activity. In addition, autonomic sympathetic activity stimulates the release of adrenaline (epinephrine) and noradrenaline (norepinephrine) from the adrenal medulla and subcortical areas in the brain. Adrenaline stimulates glycolysis, to provide glucose in the blood for energy, and both these hormones act to increase heart rate, blood pressure and cognitive abilities in the short term. Noradrenaline is also an important neurotransmitter within the central nervous system (CNS): noradrenergic cells connect the limbic system with the brainstem and the forebrain, making it integral in the initiation and recognition of the stress response.

The other arm of the neuroendocrine response is mediated through the hypothalamus, which results in the release of corticosteroids from the adrenal cortex. Cortisol has a profound effect on glucose metabolism: blood glucose levels are increased to provide a supply of energy for muscular activity through the breakdown of carbohydrates, proteins and fats. Cortisol also has a direct effect on the brain, stimulating the initiation and/or inhibition of behavioural responses. Almost every cell in the body has receptors for cortisol, which means that it has a wide-ranging effect on metabolic processes. Increased release of cortisol, for example, decreases the sensitivity of gonads to luteinizing hormone (LH) and so there is decreased release of sex hormones during the stress response. It is this effect that reduces the libido and fertility of animals suffering from chronic stress. Cortisol release is controlled by a negative feedback system involving the hypothalamus, anterior pituitary and adrenal gland. The hypothalamus secretes corticotrophin releasing factor (CRF), which stimulates the anterior pituitary to produce adrenocorticotropic hormone (ACTH), which is released into the circulation and stimulates cortisol production and release from the adrenal cortex. As this is a negative feedback system, increased levels of cortisol result in a decreased release of CRF and ACTH (see Rang *et al.* 1995).

2. THE ETHOLOGICAL APPROACH

2.1 The Origin of the Cat

In the case of the domestic cat, we can make assumptions about those aspects of an environment that are likely to compromise welfare from our knowledge of the evolutionary origin and ethology of the species. The domestic cat developed from a species (the African wild cat) that evolved as a solitary territorial hunter. Living on a sparse savannah environment, this species had to maintain a territory to survive and reproduce, and used scent as a means of orientation and communication more than visual signals. The domestication of cats occurred a relatively short period of time ago (4,000 years, or possibly 9,500 years, see Chapter 3) in evolutionary terms, and over that period reproductive activity has been controlled to a much lesser extent than in other domesticated species (Bradshaw *et al.* 1999). During the process of domestication, cats have developed the ability to live together in social groups, but only when these groups occur in particular circumstances. In the feral or farm situation, social groups are generally made up of related individuals, and the size of groups is limited by the availability of food resource (Macdonald *et al.* 2000). Because each cat hunts individually, and the number of cats in the group is matched to the amount of food, there is no need for competition for resources or 'queuing', and hence no development of hierarchical systems or signals of submission or appeasement, as occur in other social species such as the dog (but see Chapter 1 for a different viewpoint).

2.2 The Domestic Environment

Despite the intervening process of domestication, we find that those situations that most frequently cause behavioural changes in domestic cats within homes can be ascribed to characteristics that were probably present in cats prior to domestication. Problems that stem from enforced contact with other cats, and those that result from changes in the olfactory environment, or moving a cat away from an environment in which scent signals are familiar (Casey & Bradshaw unpublished data), are often explained in terms of the solitary, territorial behaviour of the ancestral species. The social situation for cats in the domestic environment is very different from that in the feral or 'natural' situation. In multi-cat households, for example, cats may not be related, and can be introduced as adults, resulting in cats that do not regard each other as part of the same social group living in close proximity; hence these cats are regarded as a threat to each others' territory and resources (see Chapter 4). This is apparent from the more amicable relationships seen in sibling pairs of cats than in more arbitrary combinations of individuals (Bradshaw & Hall 1999). The same problem occurs in areas where the population density of cats within a neighbourhood is high. In addition, competition for resources, such as food and toileting locations, is created, especially in multi-cat households if all resources are restricted to one area. Because we believe that cats do not have signals for diffusing conflict, such environments are likely to be a major cause of stress for the domestic cat.

2.3 The Cattery and Rescue Environment

Several studies have indicated that cats suffer from stress when moved into a cattery, quarantine or rescue environment (e.g. Rochlitz et al. 1998). The period of time over which signs of acute stress decline and adaptation occurs varies between individual cats and individual situations, but have been described as lasting from a few days (Smith et al. 1994), to several weeks (Kessler & Turner 1997; Rochlitz et al. 1998). The stress response is caused by the unfamiliarity of the environment, including the sudden change in scent stimuli, close olfactory and visual contact with other cats, an unfamiliar and unpredictable routine (Carlstead et al. 1993), unfamiliar human carers, and often an inability to perform species-typical coping behaviours such as hiding (Carlstead et al. 1993).

3. BEHAVIOURAL AND PHYSIOLOGICAL MEASURES OF THE INDIVIDUAL

3.1 Behavioural Measures of Welfare

Since cats are descended from a territorial and largely asocial ancestor, it is probably unsurprising that most of their more obvious signals are those associated with aggression and defence (Bradshaw & Cameron-Beaumont 2000). While at the time of performance these presumably indicate acute stress, chronic stress may be less easy to determine from changes in behaviour. In recent years, a consensus has emerged that general inhibition of a wide range of behaviour is characteristic of poor welfare in cats that are confined, for example in a cage or pen in an animal shelter (McCune 1992, 1994; Rochlitz 1999). Even cats that have been caged long-term rarely exhibit the pacing stereotypies commonly seen in wild felids in confinement (Shepherdson *et al.* 1993). Behavioural indicators of stress in domestic cats have certainly received much less attention from researchers than those of most other domesticated species.

Furthermore, individual cats vary substantially in their behavioural response to the same stressors. A shy/bold continuum has been identified in many species of wild vertebrates (e.g. Wilson *et al.* 1994). This describes variation between individuals which consistently differ in their reactions to novel objects, animals or situations, ranging from withdrawal at one extreme to approach at the other. This may underlie a substantial level of individual variation in cat behaviour (McCune 1995; Lowe & Bradshaw 2001). Part of this variation appears to be heritable (Turner *et al.*1986; McCune 1995) and may have been maintained within the cat population by its relatively undomesticated state relative to other common domestic species (Bradshaw *et al.* 1999).

The many contrasts between the environment to which its wild ancestor was adapted, and the various situations domestic cats find themselves in today, ensure that a wide range of potential stressors may trigger poor welfare. These include unresolved interactions with conspecifics; proximity to other species, including man, to which they have not been socialized; confinement in an area that is too small or otherwise inadequate to serve as a hunting territory; and an inability to cope with unpredicted changes in the physical environment. The identification of stressors by the behaviour that they trigger has received little systematic study, although such presumed connections are routinely made as part of the clinical diagnosis of behavioural disorders.

These two sources of variation, between individuals and between stressors, possibly account for the large number of behavioural measures of welfare that have appeared in the literature. Recently, attempts have been made to combine these into scales, to provide an integrated measure of subjective stress that is based on multiple indicators. These will be discussed after the individual behaviour patterns have been considered.

3.1.1 Expression of Single Behaviour Patterns

Chronic stress during confinement in a cage or pen for periods ranging from days to months has been studied more extensively than other scenarios. Under these circumstances many cats inhibit a range of essential maintenance behaviours, including feeding, grooming, urination, defaecation, and also locomotion; these tend to co-occur in the same individuals (McCune 1995; Rochlitz 1999). Under the same circumstances some cats attempt to hide, and when unobserved by humans may disrupt their physical environment (for example, by tearing up their bedding and overturning food and water bowls and litter trays). This behaviour is usually presumed to be driven by attempts to create an enclosed space in which to hide (McCune 1992, 1994). A number of studies have shown that providing cats with a suitable place to hide significantly reduces behavioural measures of stress. However, hiding and inhibition do not necessarily co-occur in the same individuals or under the same circumstances (e.g. Carlstead *et al.* 1993). Exaggerated defensive aggression towards people, for example caretakers (McCune 1992), and towards cats in group housing (Smith *et al.* 1994; van den Bos & de Cock Buning 1994; Kessler and Turner 1999a), may also be useful as indicators of poor welfare.

In group-housing, cats that are repeated targets of aggression tend to hide (van den Bos & de Cock Buning 1994); presumably the welfare of such cats is compromised because they are unable to disperse, the normal response to such aggression in free-living animals. The aggressors tend to occupy the centre of the room, and have been presumed to be 'dominant' (van den Bos & de Cock Buning 1994), although given the debate about the validity of stable social structures in cats, it is probably better to describe these cats as more 'confident' in maintaining an individual area of territory. By analogy with similar behaviours in group-housed mice (Barnard *et al.* 1994), the physiological indicators of stress in these confident cats may possibly be raised, but the cats may not be experiencing a significant decrement in welfare as a result (Barnard & Hurst 1996). Extreme vigilance also appears to be an indicator of decreased welfare in confined groups of cats constructed *ad hoc*, from individuals previously unknown to one another (Smith *et al.* 1994).

Inhibition of sleep, presumably in response to a perceived need for continuous vigilance, has also been used as an indicator of poor welfare (e.g. Rochlitz et al. 1998). It may be difficult to distinguish real sleep from 'feigned sleep' (Rochlitz et al. 1998) or 'defensive sleep' (Kessler & Turner 1997). Longer periods of Rapid Eye Movement (REM) sleep in cats resting on soft surfaces rather than hard surfaces (Crouse et al. 1995) may point to REM sleep being a useful indicator of sleep quality, and therefore possibly of improvement in welfare.

Purring, which occurs in a variety of circumstances in which the cat's emotional state may range from probable contentment to acute pain, may therefore be meaningless in terms of indicating a state of welfare, at least if taken in isolation. It has been interpreted as a contact-soliciting signal derived from juvenile behaviour (Bradshaw & Cameron-Beaumont 2000) and may therefore not be associated with any specific emotional state.

It has been suggested that the performance of play, a 'luxury' activity, is driven by a positive emotional ('motivationally affective') state (Fraser & Duncan 1998). In adult cats, play behaviour appears to be inhibited by chronic stress (Konrad & Bagshaw 1970; Carlstead et al. 1993), and the absence of play is correlated with greater behavioural scores of stress when measured by the Cat-Stress-Score (CSS), as described in the following section (Vandenbussche et al. 2002). However, the performance of play does not necessarily indicate good welfare. For example, social play, particularly 'play-fighting', may be stressful in some cases, especially in confined, artificially constituted groups. Object play is motivationally very similar, and possibly identical, to predatory behaviour, and may, for example, indicate that a cat is hungry (Hall & Bradshaw 1998), though since predation and satiety are not closely linked (Adamec 1976), fully-fed cats will also play.

Measures of acute stress have received little research attention, but van den Bos (1998) showed that self-grooming, including licking of the nose and mouth, scratching with the hind claws, and head-shaking, often occur immediately following an attack from another cat, and may therefore be displacement activities following conflict. The Global Assessment Score (see below and Table 1; McCune 1992) declines over the first 10 minutes after caging and may therefore measure some aspects of acute stress, in addition to chronic stress.

An ethogram for the domestic cat, including many of the postures and behaviour patterns referred to above, has been published (UK Cat Behaviour Working Group 1995).

3.1.2 Integrated Measures

Individual differences in the expression of welfare status, and general uncertainty over the reliability of individual measures of behaviour, suggest that at the current state of knowledge several behavioural elements taken together may provide more useful information than any single measure. A composite behavioural scale for quantifying stress in confined cats has been derived and refined, and has been widely adopted. The scale was originally devised by McCune (1992) with ten states (Table 1), later reduced to seven (McCune 1994) ranging from 'Relaxed' to 'Terror'. The latter was amplified by adding more postural elements, still retaining seven levels (Table 2), by Kessler and Turner (1997), to form the 'Cat-Stress-Score' (CSS) (Figures 2 and 3). All versions appear to be reliable between observers: McCune (1992) indicates 100% intra- and 99% inter-observer repeatability, while Kessler and Turner (1997) report 90% agreement between trained observers and 75% between untrained shelter staff.

Validation of the CSS has included a progressive reduction in the score in the days following introduction into a boarding cattery (Kessler & Turner 1997) and animal shelters (Kessler & Turner 1999a,b; Kakuma & Bradshaw 2001); a higher CSS in cats not socialized to people, caused by the proximity of the person evaluating the cat (Kessler & Turner 1999a); a higher CSS in group-housed cats not socialized to other cats compared to single-housed cats (Kessler & Turner 1999a); an increase with cat density when group-housed (Kessler & Turner 1999b). CSS has been found to be a useful measure in the assessment of enrichment devices (Kly & Casey, unpublished), and in the evaluation of the selection process by potential owners within a shelter (Vandenbussche *et al.* 2002). However, it is unclear to what extent the intervals between the seven levels are equivalent, for example whether a change from level 3 to level 2 represents an equivalent improvement in welfare as does a change from level 6 to level 5. At present, it may therefore be prudent to use non-parametric statistics (for example medians) when comparing CSS values that vary widely, though means and standard deviations have been used when comparing values that vary over only two or three levels. It is also unclear from the published versions of the test whether half-steps are permitted, for example when a cat is showing some postural and behavioural elements at level 3 and others at level 4.

In its current form, the CSS is scored by an observer who is within plain view of the cat, and the information that it gives may therefore overlap with that produced by human approach tests (see next section). Cats behave differently towards one another in the presence of a human observer than when no human is present (Nott & Bradshaw 1994), so it is possible that their CSS might be different if recorded remotely by video. Kessler &

Turner (1997) also point out that a relaxed cat's posture is affected by temperature, and do not recommend use of their scale below 15°C.

Table 1. Composite behavioural scale for quantifying stress (adapted from McCune 1992, Appendix VI, with some definitions, *in italics*, from Appendix III).

Score	Description of behaviour, posture and appearance
1	Completely relaxed, laid out body, possibly on back, pupils normal *(neither dilated, i.e. large with little iris visible, nor reduced to vertical slit)*, ears pricked forward, purring, slow blink, tail usually extended, chin may be held upwards or laid on a surface, whiskers forward or normal *(90° to centre of nose line)*, may 'greet', may drool
2	May be rubbing, purring or miaow, when approached. Ears forward, pricked; possibly at front of cage; whiskers normal or forward, slow blink. Slightly more 'aware' than in score 1, may face away from observer.
3	Belly/rear still may be exposed. Ears forward or pricked or normal (midway between forward and back). Chin may be on a surface, legs may be stretched out. Paws may be turned in, especially when approached.
4	May sit away from direction of front of cage. Whiskers normal or forward. Ears normal or forward.
5	Eyes partially dilated or normal. May miaow, explore cage, looking around. Head moves around. Chin generally not on a surface. Body a little 'tense'
6	Eyes dilated. Ears may be just slightly flattened and may be back or forward on head. Whiskers normal. 'Tense' posture. Tail may be loosely round body. May give miaow or plaintive miaow. May be actively exploring or trying to escape.
7	'Stiff' posture. Will focus on observer. Plaintive miaows or miaows. Ears towards back. Less movement of head. Pupils dilated. May try to escape.
8	Pupils dilated or very dilated. Prowling or motionless. May try to escape. Yowl. Ears somewhat flattened, back on head. Head, body somewhat crouched. Tail close to body.
9	Pupils very dilated. Close, crouched look: body close to ground, tightly bunched and usually directly on top of all fours. Tail close to body. Fast ventilation rate. May be shaking. Back of cage. Usually quiet or very vocal. May hiss on approach. Whiskers back.
10	All out defence. Hair, body and head flattened. Pupils very dilated. Tail wrapped close to body. Warning hiss or spit on approach. Far back in cage. Sits low on all fours with head low. Eyes open. May 'rage'. Very fast ventilation rate.

Table 2. Definitions of seven-level Cat-Stress-Score (from Kessler & Turner 1997). *i:* (or unspecified) = cat is inactive. *a:* = cat is active.

Score	Body	Belly	Legs	Tail	Head	Eyes	Pupils	Ears	Whiskers	Vocalisation	Activity
1 Fully relaxed	laid out on side or on back	exposed, slow ventilation	fully extended	extended or loosely wrapped	laid on surface with chin up or on the surface	closed or half opened, may be blinking slowly	normal	half-back (normal)	lateral (normal)	none	sleeping or resting
2 Weakly relaxed	*i:* laid ventrally or half on side or sitting *a:* standing or moving, back horizontal	exposed or not exposed, slow or normal ventilation	*i:* bent, hind legs may be laid out *a:* when standing, extended	*i:* extended or loosely wrapped *a:* up or loosely downwards	laid on the surface or over the body, some movement	closed, half opened or normal opened	normal	half-back or erected to front	lateral or forward	none	sleeping, resting, alert or active, may be playing
3 Weakly tense	*i:* laid ventrally or sitting *a:* standing or moving, back horizontal	not exposed, normal ventilation	*i:* bent *a:* when standing, extended	may be twitching; *i:* on the body or curved backwards, *a:* up or tense downwards	over the body, some movement	normal opened	normal	half-back or erected to front or back and forward on head	lateral or forward	miaow or quiet	resting awake, or actively exploring
4 Very tense	*i:* laid ventral, rolled or sitting *a:* standing or moving, body behind lower than in front	not exposed, normal ventilation	*i:* bent *a:* when standing, hind legs bent, in front extended	*i:* close to the body *a:* tense downwards or curled forward, may be twitching	over the body or pressed to body, little or no movement	widely open or pressed together	normal or partially dilated	erected to front or back, or back and forward on head	lateral or forward	miaow, plaintive miaow or quiet	cramped sleeping, resting or alert, may be actively exploring, trying to escape
5 Fearful, stiff	*i:* laid ventrally or sitting *a:* standing or moving, body behind lower than in front	not exposed, normal or fast ventilation	*i:* bent *a:* bent near to surface	*i:* close to the body *a:* curled forward close to the body	on the plane of the body, less or no movement	widely opened	dilated	partially flattened	lateral or forward or back	plaintive miaow, yowling, growling or quiet	alert, may be actively trying to escape

Table 2. *(continued)*. Definitions of seven-level Cat-Stress-Score (from Kessler & Turner 1997). *i:* (or unspecified) = cat is inactive. *a:* = cat is active.

Score	Body	Belly	Legs	Tail	Head	Eyes	Pupils	Ears	Whiskers	Vocalisation	Activity
6 Very fearful	*i:* laid ventrally or crouched directly on top of all paws, may be shaking *a:* whole body near to ground, crawling, may be shaking	not exposed, fast ventilation	*i:* bent *a:* bent near to surface	*i:* close to the body *a:* curled forward close to the body	near to surface, motionless	fully opened	fully dilated	fully flattened	back	plaintive miaow, yowling, growling or quiet	motionless alert or actively prowling
7 Terrorised	crouched directly on top of all fours, shaking	not exposed, fast ventilation	bent	close to the body	lower than the body, motionless	fully opened	fully dilated	fully flattened back on head	back	plaintive miaow, yowling, growling or quiet	

Figure 2. Based on its appearance, posture and behaviour, this cat has a Cat-Stress-Score of 1. (Courtesy of Yoshie Kakuma).

Figure 3. Based on its appearance, posture and behaviour, this cat has a Cat-Stress-Score of 4. (Courtesy of Yoshie Kakuma).

3.1.3 Behavioural Testing

Particular aspects of cat welfare can be evaluated by specific tests, in addition to the more general measures described above. Additions to the physical environment may be assessed by the extent to which they are used (e.g. Smith *et al.* 1994; DeMonte & le Pape 1997)), although since cats habituate rapidly to novel objects (Hall *et al.* 2002), measures should be taken over an extended period. If ethical, the effect of removal of the objects should also be assessed. The Spread of Participation statistic has been employed to assess use of space following changes in density, for wild felids in zoos (Shepherdson *et al.* 1993), and might be useful for domestic cats. Likewise, the benefits of environmental enrichment for captive felids have been successfully assessed using Shannon's Index of Behavioural Diversity (Shepherdson *et al.* 1993) but this also does not seem to have been applied to the domestic cat.

The proximity of humans and/or conspecifics are important stressors for many cats, with considerable variation between individuals which is partially affected by prior socialization (Kessler & Turner 1999a). The extent to which a particular cat is affected by such stressors may be assessed by standardised testing, usually involving the progressive introduction of a person or test cat, towards the subject cat. Kessler & Turner (1999a) briefly describe a 'Cat-Approach-Test', in which a cage containing a 'calm, neutered male cat' was placed 1 m from the subject cat and the subject's reaction noted on a six-point scale (ranging from extremely friendly to extremely unfriendly). No validation of this method has been published, other than comparison with the CSS for the same subjects when group- or single-housed (see above); for example, it would be interesting to compare the effect of test cats of different sexes, colours, and sizes. Consideration would need to be given to the welfare of the test cat if such tests were to be used routinely; models of cats may be effective if used only once on each subject (Bradshaw & Cameron-Beaumont 2000).

Testing of the stress experienced by cats due to the proximity of people has been measured in two ways, either by allowing the cat to approach a test person (McCune 1992) or by the test person approaching the cat from outside its cage (Podberscek *et al.* 1991; McCune 1992; Kessler & Turner 1999a), or holding it for a predetermined time (Lowe & Bradshaw 2002). For example, in the 'Human-Approach-Test' devised by Kessler & Turner (1999a) an observer approaches the cat's cage from the front, greets the cat verbally, stands for one minute in front of the cage touching the bars with one hand, and then opens the door of the cage for a few seconds and finally shuts it. The reaction of the cat is scored on a six-point scale (Table 3). In such tests it may be critical, for some cats at least, whether the test person is familiar or unfamiliar, although Horsfield (1998) found little inter-individual difference between reactions to familiar and unfamiliar people in approach and holding tests performed on house cats. Another study, in a population of shelter cats, found the 'Approach Test' to be neither correlated to CSS nor to time from admission to selection by a new owner (Vandenbussche et al. 2002).

Other tests have included performance in discrimination tasks (Meier & Stuart 1959), and responses to unfamiliar objects (Meier & Stuart 1959; McCune 1995).

Table 3. Human-Approach-Test scores (Kessler & Turner 1999a). The reaction of the cat to the observer's approach is recorded on a six-point scale.

Test score	Reaction of cat
1	reacts in an extremely friendly way to people
2	reacts in a friendly way to people
3	turns towards people
4	moves away or avoids any contact with people
5	reacts in an unfriendly way to people
6	reacts in an extremely unfriendly way to people

3.1.4 Abnormal Behaviour

Some, but not all, of the behavioural problems presented clinically by cat owners indicate compromised welfare (see Chapter 4). Scratching furniture, for example, is a normal part of the domestic cat's behavioural repertoire that some owners find unacceptable. Changes in elimination behaviour or the onset of indoor marking are essentially normal behavioural responses to some form of change in the cat's environment: in some cases the environmental change may also cause the animal to be stressed, but this is not necessarily the case. Over-grooming and other forms of self-mutilation, and even some ostensibly medical disorders such as idiopathic cystitis (Cameron et al. 2004), almost certainly reflect underlying stress and therefore, presumably, suffering. Behaviours such as hiding, fleeing and crouching have been classified as 'symptoms of anxiety' when performed by house cats (Heidenberger 1997), and since these are considered to be associated with stress in caged cats (see above), there is no reason to suppose that the same is not true in the domestic environment. To date there have been few attempts to map the symptoms of behavioural disorders on to behavioural measures of welfare. However, early indications of clinical studies suggest that certain patterns of problem behaviour are more closely associated with behavioural signs of stress than others. For example, cats showing change in elimination behaviour appear to show more signs of stress than those engaging in marking behaviour. Types of problem behaviour are also linked to both environmental factors and 'personality' or 'behavioural style' (Casey, unpublished).

3.2 Physiological Measures of Welfare

Physiological measures have also been used to identify the stress response in cats. Physiological measures of the stress response in individual animals have the advantage that they are quantitative: a value is obtained that can be compared with values obtained in different conditions or from other individuals. However, not only is there a huge individual variation in physiological response to stressors, but also there are normal variations at different times of day that can make comparison and interpretation difficult (Rushen 1991). Measurable physiological parameters do not co-vary, and there are often problems in measurement because animals must be habituated to sampling or measuring techniques to ensure that it is not these processes themselves that are causing changes in the stress response (Mason & Mendl 1993). In addition, some measures have the disadvantage of requiring invasive sampling methods. The stress response can be assessed through direct measurement of blood pressure, heart rate and respiratory rate but these responses are not specific to stress, as they are also elevated with exercise. However, in caged cats where exercise is not possible, ventilation rate (measured by direct observation of the abdomen) has been found to be a reproducible measure of stress (McCune 1992). More commonly, stress responses are measured through the activity of the hypothalamic-pituitary-adrenal (HPA) system, cortisol levels, or the sensitivity of the adrenal gland (via adrenocorticotropic hormone (ACTH) stimulation tests) (Klemcke 1994).

3.2.1 Cortisol

Cortisol can be measured in plasma, urine or faeces. Measurement in plasma best reflects the stress response at the time of measurement but, as well as varying throughout the day, it is also influenced by the stress response induced by handling the cat in order to obtain the blood sample. In addition, because it is invasive, sampling cannot be used for research purposes without appropriate licensing (the Home Office in the United Kingdom).

In cats, cortisol has only been measured occasionally, and usually in the urine (e.g. Carlstead *et al.* 1993; Rochlitz *et al.* 1998). Because urine is filtered out by the kidneys and collects in the bladder over a period of time, urinary cortisol gives a measurement that reflects the stress response over the 4 to 8 hour period prior to collection. This period of pooling can vary, however, particularly when cats inhibit urination during periods of acute stress, such as when first entering a rescue shelter or other novel environment. Collection of urine is relatively simple in cats, using double litter trays where the top tray contains non-absorbent litter, such as commercially produced disposable copolymeric beads or washed and sterilised gravel, and has holes drilled in it to allow urine to run through to the lower tray, preventing faecal contamination. It is usual to measure the ratio of cortisol to creatinine rather than absolute measures of cortisol, to avoid the effect of variable rates of urine production and concentration effects. Creatinine is a metabolite produced by the breakdown of muscles, and is used for this purpose because it is eliminated from the body at a relatively constant rate. Some changes in creatinine excretion do occur however with exercise, hence results should be viewed with caution where large differences in exercise rate are observed between individuals or between samples. Despite this, creatinine is still commonly used as a marker for concentration because it is the most standard metabolite that is easily measured in urine. Cortisol levels are determined by radioimmunoassay (RIA) or enzyme-linked immunosorbent assay (ELISA), and creatinine levels by spectrophotometry (Cauvin *et al.* 2003).

Measurement of faecal cortisol can also be used (Schatz & Palme 2001). The assay involves the use of a RIA for cortisol metabolites, which vary from species to species and possibly from one individual to another, and hence their results need to be interpreted with more caution. The results are likely to be less specific for measuring changes in stress within an individual, although they can be of value where an assessment is needed of the welfare of a larger group of animals in response to a change in their environment, particularly where the collection of individual urine samples is not possible. Saliva contains cortisol and can be collected less invasively than blood, but McCune (1992) could find no correlation between salivary and plasma cortisol. It is also difficult to collect a sufficient amount of saliva for analysis from most cats.

Cortisol levels are useful when measuring the welfare of large groups of animals, or looking at changes in welfare status of individuals in different environmental conditions. However, individual measures of cortisol from animals are of no value. This is because of the other factors that can influence the plasma cortisol of an individual. Cortisol is normally released in a pulsatile manner (Ladewig 1987) so levels change rapidly in the plasma. Levels also vary with diurnal rhythm, sexual behaviour (including nursing) (Walker *et al.* 1992), social activity and chronic cold.

In addition, there is such a large individual variation in cortisol production that measurement in a single animal would be meaningless. A baseline level would need to be established for such a sample to be meaningful, and even then the results may be difficult to interpret. Carlstead *et al.* (1992) found individual differences in response to mild imposed stress: in four of the eight cats tested, urinary cortisol increased, whereas in the other four it decreased. The latter appeared to have unusually high basal cortisol. Carlstead *et al.* (1992) suggested that these cats had found certain aspects of the imposed stress (mainly being handled) rewarding, accounting for the decline, if not the high basal levels. Cortisol responses can also vary with age and sex; Garnier *et al.* (1990), for example, found that bitches had a greater cortisol response to a novel environment than dogs, and levels can also vary with stages of the oestrus cycle.

3.2.2 Adrenocorticotropic hormone stimulation tests

ACTH stimulation tests can be useful as a measure of the effect of an environmental change within a research context. The adrenal cortex changes in sensitivity subsequent to chronic stimulation by ACTH (Restrepo & Armario 1987); hence, measuring the cortical response (cortisol production) to ACTH both before and after a period of chronic stress can reveal a change in response related to the degree of over-stimulation of the adrenal cortex that occurred during that period. In cats, response to ACTH stimulation has been used to show an enhanced adrenal sensitivity in cats subjected to an unpredictable caretaking routine (Carlstead *et al.* 1993). As with measuring cortisol directly, ACTH stimulation results in a large variation within a species so baseline or repeated measures are needed for results to be meaningful. Levels of stimulation in stressed animals, although increased from their baseline level, will generally still be within the normal range quoted by laboratories for that species. ACTH stimulation, as with measurement of cortisol without stimulation, can vary with sex and age (Mendl *et al.* 1992).

3.2.3 Luteinizing Hormone Releasing Hormone

Luteinizing hormone releasing hormone (LHRH) is produced by the hypothalamus to stimulate the production of lutenising hormone (LH) from the pituitary gland. It is the surge of LH that causes oestrogen release from the ovaries, which in turn initiates the positive feedback mechanism leading to ovulation. In animals that are chronically stressed, reproductive activity is depressed because cortisol reduces the sensitivity of the pituitary gland to LHRH, leading to reduced fertility or complete loss of cyclical activity. By measuring the LH response to LHRH it is possible to determine changes in sensitivity of the pituitary gland due to cortisol levels. As with ACTH stimulation, only comparative measures are of value, so baseline levels need to be established prior to implementing the environmental change that is to be evaluated. In cats, response to LHRH stimulation has also been used to show a reduced pituitary sensitivity in cats subjected to an unpredictable caretaking routine (Carlstead *et al.* 1993) but apart from this study this measure does not appear to have been widely used for cats.

3.2.4 Testosterone

Carlstead *et al.* (1993) also tested the testosterone response in male cats following injection of gonadotrophin releasing hormone (GNRH), but found no interaction with the stress of unpredictable husbandry.

4. COMPARING DIFFERENT MEASURES

The interpretation of measures of welfare presents two major problems. First, the same underlying level of stress may induce different behaviour in different cats, due to their individual 'behavioural styles' (McCune 1994; Mendl & Harcourt 2000; Lowe & Bradshaw 2002); the example of hiding and inhibition occurring in different cats has already been mentioned. Second, the extent to which particular behavioural or physiological measures reflect an underlying state of suffering is at least as unclear for this species as for any other, and so the default position is usually to give the animal the benefit of the doubt (Kennedy 1992). Since physiological measures of welfare have not yet been widely used in cats, it is still uncertain whether they are more or less reliable predictors of suffering than are behavioural measures, but they may also be affected by differences in coping style. For example, Carlstead *et al.* (1993) found a negative correlation between cortisol and hiding behaviour, suggesting that for some cats hiding may be an effective means of reducing stress. In a recent study of 23 cats newly admitted to an animal shelter, Hawkins *et al.* (2003) found a negative correlation between the rate of decline of the CSS and the rate of decline of cortisol. This implies a spectrum of coping styles, in which some cats continue to behave as if they were stressed but adapt to the unfamiliar surrounding physiologically, while others appear to become more relaxed but change little physiologically. Further validation of the CSS and other behavioural indices appears to be necessary before our understanding of the underlying processes that they measure can be complete.

5. CONCLUSIONS

Considering the popularity of the domestic cat as a pet, it is perhaps surprising that relatively little attention has been paid to validating measures of its welfare, by comparison with the amount of published material available for farm animals. There is still some uncertainty as to the interpretation of different measures of welfare, and the relative weight that should be ascribed to each. Judging by the ways in which the welfare science of other species has progressed, better integration between physiological and behavioural indicators (e.g. Mason *et al.* 2001), alternative measures to account for individual differences (Mason & Mendl 1993), and ways of addressing cognitive processes directly (e.g. Harding *et al.* 2004), may all offer ways forward. Moreover, most of the validation of welfare indicators that has been done has addressed the relatively small proportion of cats that are housed in confinement, such as those in catteries, rescue shelters, and laboratories. Very little attention has been paid to the welfare of the pet cat or feral and stray cat populations, perhaps because the assumption is often made that pet cats generally experience good, and ferals and strays generally experience poor, welfare. The clinical approach, addressing the welfare of individuals in the domestic context, and the welfare science approach, addressing the welfare of confined populations, have so far developed in parallel. Integration between them may have the potential to yield considerable insights. Furthermore, almost no attention has been paid to developing measures of good welfare ('pleasures', sensu Fraser & Duncan 1998) for the cat, and these would be valuable in assessing whether programmes aimed at improving a cat's environment have actually delivered an improvement in welfare.

6. ACKNOWLEDGEMENTS

We are grateful to Sarah Benge, Ruud van den Bos, Sandra McCune, Yoshie Kakuma, Debbie Smith, Kim Hawkins, and Sylvia Vandenbussche for valuable discussions on the interpretation of cat welfare indicators.

7. REFERENCES

Adamec, R.E. (1976) The interaction of hunger and preying in the domestic cat (*Felis catus*): an adaptive hierarchy? *Behavioural Biology* **18**, 263-272.
Ader, R. and Cohen, N. (1993) Psychoneuroimmunology: conditioning and stress. *Annual Review of Psychology* **44**, 53-85.

Barnard, C.J., Behnke, B.M. and Sewell, J. (1994) Social behaviour and susceptibility to infection in house mice (*Mus musculus*): effects of group size, aggressive behaviour and status-related hormonal responses prior to infection on resistance to *Babesia microti*. *Parasitology* **108**, 487-496.

Barnard, C.J. and Hurst, J.L. (1996) Welfare by design: the natural selection of welfare criteria. *Animal Welfare* **5**, 405-433.

Bradshaw, J.W.S. and Hall, S.L. (1999) Affiliative behaviour of related and unrelated pairs of cats in catteries: a preliminary report. *Applied Animal Behaviour Science* **63**, 251-255.

Bradshaw, J.W.S., Horsfield, G.F., Allen, J.A. and Robinson, I.H. (1999) Feral cats: their role in the population dynamics of *Felis catus*. *Applied Animal Behaviour Science* **65**, 273-283.

Bradshaw, J.W.S. and Cameron-Beaumont, C. (2000) The signalling repertoire of the domestic cat and its undomesticated relatives. In Turner, D.C. and Bateson, P. (eds.). *The Domestic Cat: the biology of its behaviour*, 2nd edn., Cambridge University Press, Cambridge, pp. 67-93.

Broom, D.M. (1988) The scientific assessment of animal welfare. *Applied Animal Behaviour Science* **20**, 5-19.

Cameron, M.E., Casey, R.A., Bradshaw, J.W.S., Waran, N.K. and Gunn-Moore, D.A. (2004) A study of the environmental and behavioural factors that may be associated with feline idiopathic cystitis. *J. Small Animal Practice* **45**, 144-147.

Carlstead, K., Brown, J.L., Monfort, S.L., Killens, R. and Wildt, D.E. (1992) Urinary monitoring of adrenal responses to psychological stressors in domestic and non-domestic felids. *Zoo Biology* **11**, 165-176.

Carlstead, K., Brown, J.L. and Strawn, W. (1993) Behavioral and physiological correlates of stress in laboratory cats. *Applied Animal Behaviour Science* **38**, 143-158.

Cauvin, A. L., Witt, A. L., Groves, E., Neiger, R., Martinez, T. & Church, D. B. (2003) The urinary corticoid:creatinine (UCCR) in healthy cats undergoing hospitalisation. *J. Feline Medicine and Surgery*, **5**, 329-333.

Crouse, S.J., Atwill, E.R., Lagana, M. and Houpt, K.A. (1995) Soft surfaces: a factor in feline psychological well-being. *Contemporary Topics in Laboratory Animal Science*, **34**, 94-97.

de Monte, M.and Le Pape, G. (1997) Behavioural effects of cage enrichment in single-caged adult cats. *Animal Welfare* **6**, 53-66.

Duncan, I.J.H. and Fraser D. (1997) Understanding animal welfare. In Appleby, M.C. and Hughes, B.O. (eds.), *Animal Welfare*, CABI Publishing, Wheathampstead, pp. 19-31.

Fraser, D. and Duncan, I.J.H. (1998) 'Pleasures', 'pains' and animal welfare: toward a natural history of affect. *Animal Welfare* **7**, 383-396.

Garnier, F,, Benoit, M., Ochoa, R. and Delatour, P. (1990) Adrenal cortical response in clinically normal dogs before and after adaptation to a housing environment. *Laboratory Animals* **24**, 40-43.

Hall, S.L. and Bradshaw, J.W.S. (1998) The influence of hunger on object play by adult domestic cats. *Applied Animal Behaviour Science* **58**, 143-150.

Hall, S.L., Bradshaw, J.W.S. and Robinson, I.H. (2002) Object play in adult domestic cats: the roles of habituation and disinhibition. *Applied Animal Behaviour Science* **79**, 263-271.

Harding, E.J., Paul, E.S. and Mendl, M. (2004) Animal behaviour – cognitive bias and affective state. *Nature* **427**, 312.

Hawkins, K.R., Bradshaw, J.W.S. and Casey, R.A. (2003) Correlating behavioural and physiological measures of stress in domestic cats in a rescue shelter. *Proceedings of the 37th International Conference of the International Society for Applied Ethology*, Abano Terme, Italy, 51.

Heidenberger, E. (1997) Housing conditions and behavioural problems of indoor cats as assessed by their owners. *Applied Animal Behaviour Science* **52**, 345-364.
Horsfield, G.F. (1998) *Behavioural Aspects of the Population Genetics of the Domestic Cat.* PhD thesis, University of Southampton.
Kakuma, Y. and Bradshaw, J.W.S. (2001) Effects of a feline facial pheromone analogue on stress in shelter cats. In Overall, K.L., Mills, D.S., Heath, S.E. and Horwitz, D. (eds.), *Proceedings of the Third International Congress on Veterinary Behavioural Medicine, Vancouver, Canada.* Universities Federation for Animal Welfare, Wheathampstead, pp. 218-220.
Kennedy, J.S. (1992) *The New Anthropomorphism.* Cambridge University Press, Cambridge, pp. 114-123.
Kessler, M.R. and Turner, D.C. (1997) Stress and adaptation of cats *(Felis silvestris catus)* housed singly, in pairs and in groups in boarding catteries. *Animal Welfare* **6**, 243-254.
Kessler, M.R. and Turner, D.C. (1999a) Socialization and stress in cats *(Felis silvestris catus)* housed singly and in groups in animal shelters. *Animal Welfare* **8**, 15-26.
Kessler, M.R. and Turner, D.C. (1999b) Effects of density and cage size on stress in domestic cats *(Felis silvestris catus)* housed in animal shelters and boarding catteries. *Animal Welfare* **8**, 259-267.
Klemcke, H.G. (1994) Responses of the porcine pituitary adrenal axis to a chronic intermittent stressor. *Domestic Animal Endocrinology* **11**, 133-149.
Konrad, K.W. and Bagshaw, M. (1970) Effect of novel stimulation on cats reared in a restricted environment. *J. Comparative and Physiological Psychology* **70**, 157-164.
Ladewig, J. (1987) Endocrine aspects of stress: evaluation of stress reactions in farm animals. In Wiepkema, P.R. and van Adrichem, P.W.M. (eds.), *The Biology of Stress in Farm Animals: An Integrated Approach.* Martinus Nijhoff, Dordrecht, pp. 13-25.
Lowe, S.E. and Bradshaw, J.W.S. (2001) Ontogeny of individuality in the domestic cat in the home environment. *Animal Behaviour* **61**, 231-237.
Lowe, S.E. and Bradshaw, J.W.S. (2002) Responses of pet cats to being held by an unfamiliar person, from weaning to three years of age. *Anthrozöos* **15**, 69-79.
Macdonald, D.W., Yamaguchi, N. and Kerby, G. (2000) Group-living in the domestic cat: its sociobiology and epidemiology. In Turner, D.C. and Bateson, P. (eds.). *The Domestic Cat: the biology of its behaviour*, 2nd edn., Cambridge University Press, Cambridge, pp. 95-118.
Mason, G. and Mendl, M. (1993) Why is there no simple way of measuring animal welfare? *Animal Welfare* **2**, 301-319.
Mason, G.J., Cooper, J. and Clareborough C. (2001) Frustrations of fur-farmed mink. *Nature* **410**, 35-36.
McCune, S. (1992) *Temperament and the Welfare of Caged Cats.* D.Phil. thesis, University of Cambridge.
McCune, S. (1994) Caged cats: avoiding problems and providing solutions. *Companion Animal Behaviour Therapy Study Group Newsletter,* No. 7, 33-40.
McCune, S. (1995) The impact of paternity and early socialisation on the development of cats' behaviour to people and novel objects. *Applied Animal Behaviour Science* **45**, 109-124.
Meier, G.W. and Stuart, J.L. (1959) Effects of handling on the physical and behavioral development of Siamese kittens. *Psychological Reports* **5**, 497-501.
Mendl, M. and Harcourt, R. (2000) Individuality in the domestic cat: origins, development and stability. In Turner, D.C. and Bateson, P. (eds.). *The Domestic Cat: the biology of its behaviour*, 2nd edn., Cambridge University Press, Cambridge, pp. 47-64.

Mendl, M., Zanella, A.J. and Broom D.M. (1992) Physiological and reproductive correlates of behavioural strategies in female domestic pigs. *Animal Behaviour* **44**, 1107-1121.

Nott, H.M.R. and Bradshaw, J.W.S. (1994) Companion animals. In Wratten, S.D. (ed.), *Video Techniques in Animal Ecology and Behaviour*. Chapman & Hall, London, pp. 145-161.

Podberscek, A.L., Blackshaw, J.K. and Beattie, A.W. (1991) The behaviour of laboratory colony cats and their reactions to a familiar and unfamiliar person. *Applied Animal Behaviour Science* **31**, 119-130.

Rang, H. P., Dale, M. M. and Ritter, J. M. (1995) *Pharmacology*. 3rd edn., Churchill Livingstone, Edinburgh, pp. 433-443.

Restrepo, C. and Armario, A. (1987) Chronic stress alters pituitary adrenal function in prepubertal male rats. *Psychoneuro-endocrinology* **12**, 393-398.

Rochlitz, I. (1999) Recommendations for the housing of cats in the home, in catteries and animal shelters, in laboratories and in veterinary surgeries. *J. Feline Medicine and Surgery* **1**, 181-191.

Rochlitz, I., Podberscek, A.L. and Broom, D.M. (1998) Welfare of cats in a quarantine cattery. *Veterinary Record* **143**, 35-39.

Rushen, J. (1991) Problems associated with the interpretation of physiological data in the assessment of animal welfare. *Applied Animal Behaviour Science* **28**, 381-386.

Schatz, S. & Palme, R. (2001) Measurement of faecal cortisol metabolites in cats and dogs: a non-invasive method for evaluating adrenocortical function. *Veterinary Research Communications*, **25**, 271-287.

Schwaber, J.S., Kapp, B.S., Higgins, G.A. and Rapp, P.R. (1982) Amygdaloid and basal forebrain direct connections with the nucleus of the solitary tract and the dorsal motor nucleus. *J. Neuroscience*, **2**, 1424-1438.

Selye, H. (1956) *The Stress of Life*. McGraw-Hill, New York.

Shepherdson, D.J., Carlstead, K., Mellen, J.D. and Seidensticker, J. (1993) The influence of food presentation on the behavior of small cats in confined environments. *Zoo Biology* **12**, 203-216.

Smith, D.F.E., Durman, K.J., Roy, D.B. and Bradshaw, J.W.S. (1994) Behavioural aspects of the welfare of rescued cats. *Journal of the Feline Advisory Bureau* **31**, 25-28.

Turner, D.C., Feaver, J., Mendl, M. and Bateson, P. (1986) Variation in domestic cat behaviour towards humans: a paternal effect. *Animal Behaviour* **34**, 1890-1892.

UK Cat Behaviour Working Group (1995). An ethogram for behavioural studies of the domestic cat (*Felis silvestris catus* L.) *Animal Welfare Research Report No. 8*. Universities Federation for Animal Welfare: Potters Bar, UK.

van den Bos, R. (1998) Post-conflict stress-response in confined group-living cats (*Felis silvestris catus*). *Applied Animal Behaviour Science* **59**, 323-330.

van den Bos, R. and De Cock Buning, T. (1994) Social behaviour of domestic cats (*Felis lybica* forma *catus* L.): a study of dominance in a group of female laboratory cats. *Ethology* **98**, 14-37.

Vandenbussche, S., Casey, R.A. and Bradshaw, J.W.S. (2002) Factors influencing the selection of domestic cats from a rescue shelter. *Proceedings of the BSAVA Congress, 2002, Birmingham, UK*. BSAVA Publications, Gloucester, 610.

Walker, C.D., Lightman, S.L., Steel, M.K. and Dallaman, M.F. (1992) Suckling is a persistent stimulus to the adreno-cortical system of the rat. *Endocrinology* **130**, 115-125.

Weipkema, P.R. and Koolhaas, J.M. (1992) The emotional brain. *Animal Welfare* **1**, 13-18.

Wilson, D.S., Clark, A.B., Coleman, K. & Dearstyne, T. (1994) Shyness and boldness in humans and other animals. *Trends in Ecology & Evolution* **9**, 442-446.

Chapter 3

THE HUMAN-CAT RELATIONSHIP

Penny L. Bernstein
Biological Sciences, Kent State University Stark Campus, Canton, OH 44720 USA

Abstract: With an estimated 76 million pet cats in the United States and 200 million worldwide, there is an increasing interest in, and need to understand more about, the human-cat relationship. This chapter presents the growing body of research that evaluates this relationship from a variety of perspectives. It considers the history and importance of animals as companions, worldwide trends in pet ownership, physiological and psychosocial health benefits of pet ownership, the role of pets in families and their special role in the lives of children, and the difficulties people have in dealing with the loss of their animal companions. Particular aspects of the human-cat relationship are also considered, ranging from cat socialization and the effects of paternity and breed on social behaviour, through to observational studies of human-cat interactions in the home, including cat vocalizations, petting, and social interactions, both between cats and between cats and humans. Responsibilities of pet ownership are examined, including providing veterinary health care for the animals, and minimizing zoonotic disease and other health risks to humans. Failures of the human-cat relationship can also occur, and a number of examples are considered: animal abuse and animal hoarding, feline behaviour problems, pet relinquishment and abandonment and the growing problem of free-roaming, stray and feral cats.

1. ANIMALS AS COMPANIONS

1.1 Introduction

Animals have been companions to humans since ancient times. Egyptians are often given credit for first domesticating African wild cats approximately 4,000 years ago; recent findings suggest cats may have been closely associated with humans as long as 9,500 years ago (Vigne *et al.* 2004). Ancient writings and historical records, as well as more recent studies, document the various ways in which animals and humans have related in a positive way. Animals have served, for example, as protectors (dogs protecting against bears and wolves, and cats protecting crops by killing rodents); as food providers and hunting partners; and as an important "other" for humans to interact with and talk to, in both home and institutional settings (e.g. Beck & Katcher 1996; and see Hart 1990, 2000a for an overview of psychosocial benefits). Indeed, as Beck and Katcher (1996) note in their overview of human-animal interactions and animal companionship, the word "companionship" derives from the Latin for "together" and "bread", and "eating together" is literally one of the many ways in which humans and companion animals interact, with humans giving carrots and sugar cubes to horses, choice table scraps to dogs and cats, fruit treats to parrots and so on.

1.2 Scale of Pet Ownership

At some point, animals went from being utilitarian companions to "pets", although it is not clear what the crossover involves (see a series of discussions beginning with Eddy 2003a and followed by Copeland 2003; Eddy 2003b; Hart 2003; Lawrence 2003; Rollin & Rollin 2003; and Sanders 2003). That this "pet" companionship has been successful for both parties, however, is suggested by the dramatic increase in the population of companion animals. In 2002 in the United States it was estimated that there were between 55 and 61 million pet dogs and nearly 76 million pet cats (Euromonitor International 2003).

Such trends are carefully tracked worldwide by the pet food and pet care industries. A recent market study by Euromonitor International (2003) provides detailed information about pet ownership and notes key elements affecting pet numbers internationally. As might be expected, countries with larger populations tend to have more pets, demonstrated most obviously by China and the United States, ranked first and third in human population and second and first in pet ownership, respectively (Table 1). However, this relationship does not always hold: India is second in human population but a distant 29th in terms of pet ownership, while just four million Australian households collectively care for over 26 million pets. Demographic trends, such as increases in single households and urbanization, apparently favour ownership of smaller, easy-to-care-for animals such as cats, fish, birds, and even rabbits and ferrets, over larger, more care-intensive animals such as dogs.

The Euromonitor International report (2003) also suggests that population attitudes can affect ownership; there is little growth where dogs and cats are still regarded primarily as working animals and often rapid growth where they are seen more as companions. This supports findings documented by Serpell (1985) in a review of the anthropological literature up to that time. This trend is most evident in Turkey, Brazil, China and Thailand, where increases in disposable income, coupled with an attitude shift and increasing urbanization, has resulted in dramatic increases in pet ownership. Turkey showed a 39% increase and Brazil a 28% increase in pet ownership from 1998 to 2002, and Thailand's pet population nearly doubled in this time period (reaching 14 million in 2002), resulting in that country becoming the fifth largest in pet ownership in the Asia-Pacific area. China showed the greatest numerical increase in its pet population over this period, expanding by nearly 40 million pets.

Table 1. Pet numbers for the top 15 pet-keeping countries (arranged in order by total number of pets). Data for 2002; numbers are in millions (Euromonitor International 2003).

Country	Number of pets	Cats	Dogs	Other
United States	366,370	76,430	61,080	228,860
China	271,774	53,100	22,908	195,766
Japan	75,372	7,300	9,650	58,422
Germany	72,600	7,000	Not in top 15	61,300
Brazil	67,005	12,466	30,051	24,488
Italy	63,100	9,400	7,600	46,100
France	56,000	9,600	8,150	38,250
Russia	50,790	12,700	9,600	28,490
United Kingdom	46,590	7,700	5,800	33,090
Australia	26,625	Not in top 15	Not in top 15	21,005
Canada	22,558	6,811	Not in top 15	11,416
Turkey	22,547	Not in top 15	Not in top 15	19,208
Poland	21,315	5,465	7,520	Not in top 15
Spain	20,519	3,191	Not in top 15	13,358
Ukraine	17,635	7,350	5,425	Not in top 15

Age has a major effect on ownership as well: eight of the ten countries with the greatest number of pets have the largest populations of persons over 65 years of age (Euromonitor International 2003). This finding is in line with results of various American studies that have demonstrated positive effects of pets on attitudes and health in the elderly (e.g. Baun & McCabe 2000; Enders-Slegers 2000; Friedmann 2000; Friedmann et al. 2000). Surprisingly, personal disposable income does not seem to have a strong effect on pet ownership, according to this report. While some economically-developed countries are among the top ten, countries such as Brazil and China where owners have considerably less disposable income, also rank in this group.

Cats continue to outrank dogs as pets in the United States and China (Table 1), as well as in several other countries with much smaller numbers of pet owners (Euromonitor International 2003). Reasons cited by the report for high numbers of cats as pets include increasing urbanization (e.g. parts of

Brazil), increases in the elderly population, restrictions on dog ownership (e.g. high license fees and other restrictions), and decreases in birth rates (e.g. all three factors cited for China). However, dogs continue to outrank cats in many countries, especially where population densities are low (e.g. parts of Brazil), where crime rates are high (e.g. South Africa), or where hunting is very popular (e.g. Italy). It should be noted, however, that "other" pets, including fish, birds, rabbits, other small mammals, and reptiles, greatly outnumber both dogs and cats in most of the key market countries. Turkey and Australia (which has seen a steep decline in cat populations since 1993, see Rochlitz 2000) are extreme examples of this pattern, with over 80% of their pet population numbers coming from the "other" category.

1.3 Benefits for Humans and Companion Animals

For a number of years, studies have suggested that human and animal companions benefit from one another. Human benefits range from physiological (blood pressure control, relaxation effects, decreased levels of chemicals associated with anxiety, improved survival and longevity after heart attacks; see Friedmann 2000 and Friedmann *et al.* 2000 for overviews) to psychological (decreased depression, elevated mood or decrease of poor mood; e.g. Zasloff & Kidd 1994a,b; Turner *et al.* 2003) to practical (serving as guides, "alerters" to unwanted psychological incidents for owners with mental conditions, or in other therapeutic roles in Animal-Assisted Therapy; e.g. Beck 2000; Fine 2000b). Not all studies show benefits to humans, however. This seems to depend in part on the age of participants and what health aspects are measured (e.g. Parslow & Jorm 2003).

Benefits for animals include reliable food supplies, veterinary care, protection from disease and predators, and good environmental conditions provided by humans, as well as decreased stress levels (Carlstead *et al.* 1993). Other ways in which animals and humans may gain health benefits from one another's companionship are summarized in Fine (2000a) and Podberscek *et al.* (2000).

In addition to providing health benefits, companion animals can serve as focal points around which everyday interactions between humans can take place (e.g. Hunt & Hart 1992). They may also be an avenue for discussion of important issues such as animal consciousness and animal rights, as humans consider the problems surrounding abandoned and feral animals (e.g. Ash & Adams 2003), animal abuse (e.g. Arluke 1997; Ascione and Arkow 1999), animal hoarding (e.g. Patronek 1999; Arluke *et al.* 2002), relinquishment of animals to shelters and animal control facilities (e.g. Patronek *et al.* 1996; Salman *et al.* 1998; Scarlett *et al.* 2002) and success or failure of adoption of animals from shelters (e.g. Neidhart & Boyd 2002).

1.4 Companion Animals and the Family

Companion animals clearly play a role within the family. Triebenbacher (2000) provides a well-organized overview of the various roles, functions and contributions of pets and how those relationships and values change as families undergo life cycle changes (e.g. from having small children through children leaving the home and adults becoming elderly). She also examines problems and responsibilities associated with having companion animals, and the important effects of companion animal loss on the family. Cohen (2002) more directly examined what people mean when they say "My pet is a member of the family." In her study, pets seemed to occupy an overlapping but different space from humans. People identified pets as family members by the way in which pets functioned within the household. In response to forced-choice questions, respondents often put pets ahead of humans when making decisions (for example, to save someone when a boat tips over or to provide needed medication). Humans even celebrate special occasions with their animal companions as if they were human family members, for example, having a bar-mitzvah ceremony for a beloved horse turning 13 years of age, or a bat-mitzvah for an adored pet female cat or a dog wedding between two cherished pets (Dresser 2000).

These studies suggest that separation from pets and concern for their well-being could be a major problem for owners when they are ill, hospitalized, or taking extended trips away from home. Allowing pets to travel with their owners or visit them in hospitals, or providing pets for residents in long-term care facilities, may have more profound and important effects than previously thought. Bernstein *et al.* (2000) noted that some residents in a long-term facility preferred animal visitations to non-animal activities such as Arts and Crafts because they had had pets before entering the facility and missed those interactions, or had had pets when they were young and visits from the animals reminded them of these happier times. A number of articles in Fine (2000a) provide guidelines for developing and running visitation or Animal-Assisted Therapy programs (AAT) in a variety of settings, as does the website (www.deltasociety.org) of the Delta Society, one of the major organizations in the United States devoted to providing information and guidelines about AAT and to organizing effective programs.

1.5 Companion Animals and Children

The relationship between children and animals has received special attention from a number of researchers. Classic works include those of Levinson (1969) and Myers (1999). More recently, Triebenbacher (2000) provided an overview of the changing relationships between children and animals from early childhood through adolescence. She noted a number of skills and values children may gain from these interactions, including learning about mutual respect, kindness, humane treatment of others, giving and receiving love and affection, caretaking skills, responsibility, and pain of loss. Gail Melson has examined many of these issues in depth (e.g. Melson 2000, 2001). In her book *Why the Wild Things Are* (2001), she explains how a casual observation of a young boy, emotionally involved with his dog in a veterinarian's waiting room, had caused her to realize there was a critical lack of sociological research on child-animal interactions despite years of careful, thorough analyses of children's bonds with one another, with parents and other family members, and with humans outside of the family. Her long-term studies since then, of the relationship between children and animals, have led her to propose a view of development that recognizes the pervasiveness and importance of real and symbolic animals in children's lives. She outlines this "biocentric" approach in her book, mapping where our knowledge currently rests and where future research should be directed.

Several studies have focused on the use of animals in therapeutic programs for children. For example, Katcher and Wilkins (2000) examined therapies that employ caring for animals and nature study to help children with behaviour problems. Mallon *et al.* (2000) discuss how such programs can be developed and present the Green Chimneys model, based on that residential treatment center. The center specializes in the care of children with emotional and behavioural needs, and interaction with animals plays a significant role in therapy. Fine (2000b) discusses incorporating animals in psychotherapy programs, especially those for children.

Problems associated with children and abuse have also been studied. Ascione *et al.* (2000) examine the interrelationship between "animal maltreatment and interpersonal violence". Issues concerning animal abuse and its importance to children are examined in several ways: how children are affected by viewing animal abuse, connected or not with abuse of human family members, and stages children and adolescents go through that may or may not indicate abnormal behaviour with respect to animals and abuse.

1.6 Companion Animal Death

Owners also suffer upon the death of their pets. Books and articles point out important aspects of this part of pet ownership (e.g. Stewart 1999; Swabe 2000; Davis *et al.* 2003 for recent overviews). While natural death is clearly something owners must cope with, an additional issue is euthanasia. While technical considerations (when and how) are important, other areas that have received attention are the differing perspectives of the veterinarian, who deals with the issue frequently, versus the owner, who may have little experience with this event and therefore less ability to take a "long view" (see Stewart 1999 and as summarized by Swabe 2000), and the problem of veterinarians having to deal with owners who request euthanasia for healthy pets.

Grief is another area that has been examined, including the difficulties many owners face of feeling "silly" for grieving for a "mere" pet. Davis *et al.* (2003) sought to examine what factors, including religion and approaches to euthanasia, might help people best cope with pet death. They found, as might be expected but is rarely quantified, that all but two of 68 people in their Australian sample group reported being "sad" at the death of their pet, and that over 40% described themselves as "devastated." Religious beliefs did not seem to protect individuals from initially intense grief responses, but did seem to provide comfort over time, mostly due to beliefs in some sort of afterlife for their pet. Indeed, even participants who described themselves as atheists took comfort in this idea. Euthanasia decisions and experiences caused great conflict and distress, and having some sense of control seemed to be helpful. Veterinarians who provided clear information and clearly outlined options were identified as most helpful to clients; people also appreciated having a say in what happened to their pets.

But perhaps most important in this study was the finding that people wanted and needed to talk about their loss. The authors noted that for many owners, the pet they lost was in fact the very individual in whom they would most likely have confided about this difficult situation, and there might not be anyone else who could immediately fill that void. As has been noted in other studies, some participants were also reluctant to let others know how they were feeling "for fear of ridicule." The authors also considered special situations, such as the unique difficulties faced by disabled persons at the loss of their therapy animals. These individuals often have particularly close bonds with their companion animals, as well as a special reliance on them. Based on their findings, the authors were able to provide veterinarians with a number of specific recommendations to help their clients cope with pet loss.

1.7   Ethical Treatment of Animals

A growing animal rights movement also reflects changes in the human-animal relationship. Since 1980, with the founding of People for the Ethical Treatment of Animals, PETA (www.peta.org), modern groups such as PSYETA, Psychologists for the Ethical Treatment of Animals (www.psyeta.org) and Ethologists for the Ethical Treatment of Animals /Citizens for Responsible Animal Behaviour Studies (EETA/CRABS, www.ethologicalethics.org) have joined older groups, such as the American Anti-Vivisection Society (founded in 1883, www.aavs.org), Humane Society of America (HSUS, www.hsus.org), and the American Society for the Prevention of Cruelty to Animals (ASPCA, www.aspca.org), to name a few, to increase public awareness, change opinions, and change practices related to a variety of issues concerning both wild and companion animals. The use of dogs, cats, and even rats, mice and primates in medical, veterinary, and commercial research or for training of students has been questioned, and in many cases greatly reduced as a result of these efforts.

The United States Department of Agriculture tracks the use of animals in research. Its latest report, citing data from 2002, indicates that cat research populations have decreased from a high in 1974 of 74,000 to a low of about 22,000 to 24,000, that has been relatively stable since 1998 (www.aphis.usda.gov/ac/publications/html). Alternatives to using animals in research and training continue to be actively pursued (e.g. Bekoff *et al.* 1992; Greenfield *et al.* 1995; Hart 1998). The University of California Center for Animal Alternatives has taken a leadership position in this area, providing publications, coursework, and a web-based tool (www.vetmed.ucdavis.edu/Animal_Alternatives/main.htm) designed to improve access to a wealth of information on alternatives, especially for use in education at all levels, e.g. for school children, college, and veterinary students (Hart *et al.* 2004; Hart & Wood 2004).

1.8   The Companion Animal Bond

It is clear that companion animals are important to humans. Serpell (2002) and others (Bahlig-Pieren & Turner 1999; Rajecki *et al.* 1999; Morris *et al.* 2000) have suggested that anthropomorphism itself, "attribution of human mental states (thoughts, feeling, motivations and beliefs) to nonhuman animals," may have evolved to enable humans to recognize animals as "alternative sources of social support." This human trait may have, in turn, acted as a selective pressure on the animals themselves, "favouring physical and behavioural traits that facilitate the attribution of human mental states to nonhumans" (Serpell 2002). Some of this may involve neotenization, the retention of juvenile characteristics into adulthood, such that adult animals continue to look and act young and "baby-like". Several studies have made such claims for dogs (e.g. Coppinger & Schneider 1995; Goodwin *et al.* 1997; Coppinger & Coppinger 1998; McGreevy & Nicholas 1999). Cats may be similarly favoured because of their soft fur, small size, and the willingness of most of them to be held, petted and cuddled by humans (Figure 1). The personalities of pets and people may affect their relationship, as well (e.g. Podberscek & Gosling 2000), but understanding of this phenomenon has been hampered by difficulties of definition, cross-species comparisons, and differences in methodology.

Figure 1. Cats may be favoured as companion animals because of their soft fur, relatively small size, and the willingness of most of them to be held, petted and cuddled by humans.

Archer (1997) explored these issues in some detail in his examination of the possible evolutionary aspects of pet-keeping. In essence, "pet ownership poses a problem, since attachment and devoting resources to another species are, in theory, fitness-reducing." Like other forms of interspecific association, he argues, pet-keeping must be examined in terms of benefits and costs for the participating species. He suggests that pets evolved ways to elicit care-giving from their human partners, manipulating human responses that were originally selected to facilitate human-human interactions. The human partner in turn may be rewarded with a relationship that has fewer conditions and expectations placed upon it than those involving other humans.

Such evolutionary selection may have both positive and negative consequences. Serpell (2002) suggests, for example, that the loyalty and fidelity exhibited by many dog breeds may be a result of this process and is a positive outcome for both humans and dogs. Alger and Alger (1997) further suggest that humans and companion animals have formed shared rituals that are co-understood and performed, allowing for routine interactions. On the negative side, Serpell (2002) and others (e.g. McGreevy & Nicholas 1999) note that the breeding of dogs for specific physical traits has resulted in numerous health problems, some lethal (for example, severe sleep apnoea in English Bulldogs). Serpell (2002) also suggests that selection may have resulted in animals that are themselves so dependent on their human attachments that various behaviour problems develop as a result (such as "separation anxiety"), often leading to relinquishment of the pet. Cats have not been subjected to as much selective breeding as dogs; however, increasing interest in "tailored" cats may have effects in the future. For example, the rise in interest in Pixie cats or Munchkins, who have exceptionally short legs, or Sphinxes, that lack normal fur, may lead to problems in cats similar to those found in dogs (see Chapter 10).

Evidence has mounted, then, that the companion animal bond "must be looked upon as a kind of relationship that supplements and augments human relationships – the bond distinctively different from human relationships" (Katcher 1981), and that there are both good and bad aspects to this relationship.

2. CATS AS COMPANION ANIMALS

2.1 Scale of Cat Ownership

In some countries, cats have surpassed dogs as the most numerous companion animal, with an estimated 76 million in American homes (Euromonitor International 2003). While the United States is the world leader in cat ownership, 12 of 15 countries considered key markets for cat food and cat care products have also shown increases in cat ownership from 1998 to 2002, ranging from a modest 1.4% (Ukraine) to as much as 28.4% (Brazil) (Euromonitor International 2003). Reasons cited for these increases include cats being easy to care for (they can use a litter box so don't need to be taken outdoors), small enough to be kept easily in smaller living spaces such as apartments, and able to endure long separations without apparent problems. These traits make them ideal companions for present-day owners who work long hours, may postpone house buying until later in life, and move frequently, often to rented apartments rather than to owned houses.

2.2 Benefits of Cat Ownership

A number of factors that affect pet ownership were brought to the fore and considered in Podberscek *et al.*'s collection of papers (2000). However, cats do not play a big role in most of the studies presented. Indeed, cats form only a small part of many studies that examine favourable attributes or positive benefits of pet ownership, or are not included at all. If cats are included, they are generally ranked lower or less positively or with more mixed results than dogs. For example, Bonas *et al.* (2000, and see Serpell 2002 for overview) used a survey instrument (the Network of Relationships Inventory) to enable people to describe and evaluate the different kinds of social support they derive from both human and non-human animal relationships. Although humans scored highest overall in the aggregate, pet dogs scored higher than humans in some areas while cats ranked lower than dogs overall (although they did rank higher than other pets and scored almost as well as humans in some categories).

Enders-Slegers (2000) investigated the importance of companion animals to elderly cat and dog owners, most of whom lived alone. Her findings provided support for the idea that companion animals do play important social roles; however, only 14 of 60 participants owned cats, with an additional six owning both a cat and a dog. More than half of the participants owned only dogs.

Most physiological studies that show human health benefits focus on dogs, and when cats are included results are often mixed (see reviews by Friedmann 2000; Friedmann *et al.* 2000). For example, in a study of angina and other cardiovascular disease (Rajack 1997), results suggested that cat owners were more likely to be readmitted to hospital for further cardiac problems or angina than people who did not own pets. In addition, although people who adopted pets, either dogs or cats, from shelters experienced decreases in minor health problems one month after adopting, this effect lasted for the full 10 months of the study for dog owners but not for cat owners (Serpell 1991). Possible confounding factors in these physiological studies, however, include the fact that women were much more likely to be cat owners than men were, and were more likely to die from their heart conditions by the time they were hospitalized than were men; also physiological severity of illness was not always controlled for (Friedmann *et al.* 2000).

Studies of attachment and of the effects of cats on mood also provide mixed results. Zasloff and Kidd (1994b) examined 148 adult female students and found that having a pet could help decrease feelings of loneliness, particularly for women living alone, and compensate for the absence of human companionship. Cats and dogs seemed equally good at providing companionship. However, women living with only a cat were less attached to the cat than women living with both a cat and other people. In contrast, single women with dogs were more attached to the dog than those in multi-person households. The authors suggested that being alone with a dog allowed for more walks, rides, playing of games, travel, and other activities together, during which the dog served as a meaningful companion. Being alone with a cat does not necessarily result in any more behavioural interaction or different kinds of interaction than being with a cat in a multi-person home.

Zasloff and Kidd (1994a) examined various aspects of attachment to cats in more detail and found a more positive result. They surveyed 100 adult cat owners who seemed to be strongly attached to cats. Participants owned 267 cats, and stated that they preferred cats because of the ease of care, the affection and companionship provided by them, and their personalities. They also liked the behaviour and appearance of cats, said they felt comforted by cats, or that they simply loved cats or had "always had cats". One person said that "purring creatures who sit in your lap tend to reduce stress levels." Respondents did not like some aspects of cat behaviour: annoying behaviour (which was not further defined), lack of social behaviour or affection, destructive or aggressive behaviour, shedding and hairballs, fights with other pets, or feeding problems. The authors then compared how people felt about their cats versus humans. For example, they compared rankings (1 = strongly disagree to 4 = strongly agree) on such statements as "My cat makes me feel safe" versus "My companion makes me feel safe". They found that cats were ranked as being better than humans at making people feel needed, and providing companionship, something to care for, and something to watch. They were ranked as being worse than humans at making the participants feel safe or providing them with exercise.

Zasloff herself, however, cautions us about interpreting studies of attachment. She notes in a later study (1996) that the kind of survey questions used to assess attachment often include those specific to interactions with dogs, leading to higher attachment scores for dog owners than for owners of other pets. She developed a scale to assess this issue, based on postulates by Beck and Katcher, discussion with other researchers and other scales, primarily the Lexington Attachment to Pets Scale (Johnson et al. 1992). Using the Comfort from Companion Animals Scale (CCAS) which measures attachment in terms of comfort received from a pet, she documented that there were no differences in attachment scores between cat and dog owners on "11 items pertaining to the emotional nature of the relationship"; if two items pertaining specifically to dogs were included (relating to exercise and safety), dog owners showed a significantly higher degree of attachment.

Using an in-home interview approach, Miller and Lago (1990) sought to actually document differences in how dogs and cats interacted with owners. Again cats were underrepresented, with only 15 of 46 elderly women participants being cat owners. As might be expected, dogs inserted themselves much more into the interviews, interacting with owners and interviewers. They also were more likely to show coordinated movements with owners (e.g. owner got up, dog got up, owner sat back down, and so did the dog), make noises (such as whining, barking) and be given orders by the owner (e.g. "sit down"). Cats engaged in less social behaviour toward interviewers, and were more likely to be described by the interviewer as being calm, dignified, aloof or ignoring during the session. The main cat-related behaviours noted by interviewers were that cats let owners pick them up, the cats made friendly approaches to the interviewer (although fewer cats than dogs did this), and the owners told more stories about their cats than about their dogs. Clearly there are important differences in how cats and dogs serve as companions. While these differences may be familiar to owners, there is little scholarly research that examines these differences directly.

Albert and Anderson (1997) examined the contributions of dogs and cats to morale maintenance and positive social interaction within the family by surveying 85 families on 14 attitude items (such as "our pet helps family members cope with the normal stress of everyday life," or "our pet can sense when a family member is ill or upset."). Similar numbers of households had at least one dog or one cat. On the nine items that tested as significant, owners demonstrated that they perceived both cats and dogs as contributing to the morale of the family, that is, help "ameliorate daily stress, cope with life crises, and facilitate positive social interaction". However, dogs were perceived by owners to have a more profound impact on promoting morale and positive interaction, as demonstrated by the fact that pet type (dog) was the only variable found to have a significant effect on people's responses.

A ray of optimism about cats as companions comes from studies that focus directly on observations of the cat-human interaction, such as those by Turner and his colleagues. For example, in a number of studies this group examined how the presence of, and interactions with, a cat affected owners' moods (Rieger & Turner 1999; Turner & Rieger 2001; Turner et al. 2003). They found cats could decrease human negative moods, although the cat did not put an owner in a good mood, unlike studies of owners who were petting their dogs. Stammbach and Turner (1999) found that cats may substitute for persons in the social network or provide an additional source of emotional support, especially for those who were strongly attached to their animals. Turner (2000a) also found that cats "appear to take on the role of a significant partner in relationships involving [older] people living alone." An overview of human-cat attachment, factors that affect the relationship, and various mechanisms that seem to play a role in establishing and maintaining good relationships between cats and their owners are summarized in Turner (2000b).

Hart (2000b) also provides a positive overview of the value of cats as companions in her chapter on the selection of animals for Animal-Assisted Therapy. She stresses the flexible nature of cats, including their ability to be socially self-sufficient, to be left alone for longer periods than dogs, and to be more likely to accept care from strangers (such as neighbours or friends) if owners become ill or hospitalized. Cats also require less effort and are less demanding than dogs and require less vigorous interaction, a positive aspect for owners who are older, disabled, or fatigued from caring for others who are ill. In the same volume, Granger and Kogan (2000) outline how cats can best be used for therapy. For example, they are especially useful for those individuals who are afraid of or allergic to dogs (though cat allergies are more common), and the best cats are those who are very friendly and seek and respond well to petting and being held on a person's lap. Bernstein *et al.* (2000) showed that cats and dogs brought to clients in long-term care facilities as part of a therapy regimen enabled clients to participate in more frequent and longer lasting conversations with both humans and the animals, increase their initiation of social behaviours (both conversations and petting the animals), and dramatically increase the amount of touch in which they were involved, due primarily to petting the animals. Being able to initiate interactions and touch something outside of themselves are both important ways these patients can maintain contact with their external surroundings.

2.3   Dislike and Fear of Cats

Not all humans like cats, and some fear them. Serpell (2000) notes that throughout medieval Europe cats were seen as "malevolent demons, agents of the Devil" and associated with witchcraft; traditional Japanese folklore also depicted cats as demons. Citing Ritvo (1985), Serpell notes that even in the professional zoological literature of the 19th century, cats were the most "frequently and energetically vilified of all domestic animals." Even in the twentieth century, owners in many countries continued to voice concerns that cats would sleep on children's faces, smothering them, or "steal their breath", or cause asthma (Serpell 2000). While cats are popular pets today, they continue to also be targets of dislike.

Fear of cats may actually result in phobia, "persistent and irrational fears of a specific object, activity, or situation that is excessive and unreasonable, given the reality of the threat" (American Psychological Association, glossary at www.psychologymatters.org). Felinophobia, or ailurophobia, is one of several anxiety-related disorders now defined as "specific phobias", and is similar to agoraphobia (fear of crowds), acrophobia (fear of heights), and arachnophobia (fear of spiders). Cat phobia does not seem to be linked to a dislike of cats; people are simply irrationally afraid of them. The fear may be the result of a bad interaction with an animal in early childhood but more often the origin of a specific phobia is unknown (McNally & Steketee 1985). Although many children may fear animals at some point early in their lives, most apparently lose that fear as they grow older; adults with specific animal phobias are thought to be those whose childhood fear has failed to dissipate (McNally & Steketee 1985). Like most phobias, this fear may cause simple symptoms such as dizziness, sweating or breathlessness, or more complex and frightening reactions including heart palpitations, fear of dying, fear of becoming mad or losing control, or a full-blown panic attack. Treatments range from traditional psychotherapies to various behaviour therapies to hypnosis.

3. THE HUMAN-CAT RELATIONSHIP

3.1 Socialization and Paternity

Kittens spend much of their first few weeks with their mother, experiencing a warm, encircling "hollow", with their bodies in contact with warm surfaces (mother and litter mates). This may influence what "spots" cats prefer as adults (warm encircling areas such as pillows, laps, chair corners) and their social skills, which they are gaining at the same time. The latter develop as the kittens learn to adjust their suckling behaviour with respect to changes in the behaviour of littermates and their mother (Rosenblatt et al. 1961).

The important socialization period of cats to people seems to occur between 2 and 8 weeks (e.g. Karsh & Turner 1988) and socialization to people seems to be less effective if delayed until after 7 weeks. Several studies have addressed natural temperaments, that is behaviour expressed as kittens before extensive exposure to outside influences, while others have examined individuality, socialization to other cats, and possible ways to encourage socialization to humans, usually through increased human handling of kittens at an early age (Moelk 1944; Meier 1961; Collard 1967; Moelk 1979; Adamec et al. 1983; Meier & Turner 1985; Feaver et al. 1986; Mertens & Turner 1988; as reviewed in Mendl & Harcourt 1988 and updated in Mendl & Harcourt 2000).

Feaver et al. (1986) identified a number of cat personality types from observed behaviors, condensing them into active/aggressive, timid/nervous, and confident/easy-going. Other studies have used similar terms (e.g. Moelk 1979; Karsh & Turner 1988). Lowe and Bradshaw (2001) looked for stable personality traits by studying the post-meal behaviour of kittens and described several sets of traits that seemed to remain more or less stable for the duration of the study (4 to 24 months). The most stable behaviours were those that started as kitten investigatory behaviours (such as "inspecting the experimenter") and kitten approach behaviours to the observer (described as "boldness"). Staying indoors after eating and various rubbing-associated behaviours (such as rubbing objects or the observer, and flicking the tail) were somewhat less stable or seemed related to specific environmental or developmental conditions (such as a cat needing to urinate outside soon after eating). Several studies have demonstrated that handling by humans during early development can help kittens be less fearful and friendlier toward people, especially when contact includes talking (summarized in Karsh 1984). However, they also reveal that some kittens seem resistant to change in their original types; in other words, some friendly kittens remain friendly whether handled or not, and some fearful kittens remain so despite handling.

Some studies have suggested that the genes of the father, genetic and behavioural influences of the mother, the presence or absence of the mother during early encounters with humans, aspects of kitten curiosity and fear, and human stroking and feeding may all be factors that affect kitten responses to humans (e.g. Collard 1967; Turner et al. 1986; Reisner et al. 1994; McCune 1995; and summary in Turner 2000b). For example, Reisner et al. (1994), sought to determine if handling at early ages or other factors, such as paternity (Turner et al. 1986), would help kittens in a research colony better respond to necessary laboratory or veterinary procedures (vaccinations, etc). They found that a limited period of handling (15 minutes, three times per week for three weeks, which they felt would be a realistic protocol for assistants to follow if instituted as policy) had no clear effect on later behaviour (up to 20 weeks). Rather, paternity and litter of birth had the greatest influence on whether or not kittens were calm during procedures.

McCune (1995) sought to extend Turner's work (Turner et al. 1986), simultaneously investigating the influences of early kitten socialization to humans and friendliness of the kitten's father on subsequent interactions of kittens with humans. Each of these aspects was found to influence the behaviour of cats at one year of age. Kittens were handled or not from 2 to 12 weeks and then tested at one year of age. Testing involved three different situations per cat: a familiar person entering the test enclosure, sitting for ten minutes, and then approaching the cat; an unfamiliar person doing the same sequence; or the placing of a novel wooden box in the center of the enclosure. There was a significant effect for some behaviours, such as latency to approach within 50 cm of the test person, touch, and rub the person, and short latency correlated with the friendly father and socialized condition. Socialized one year olds from a friendly father were more likely to show relaxed behaviour and less likely to show defensive behaviour to a stranger (such as hissing, body flattening, or hiding). Socialization did not seem to play a role in responses to the novel test box; the box initially distressed all the cats. However, friendly-father cats were quicker to approach, touch, enter the box, and stay near it than were cats with unfriendly fathers. This response, based more on father characteristics than early handling, suggested that the cats had inherited or developed a "boldness" to approaching people or objects, rather than simply an increased "friendliness" toward people. Although only two fathers were involved, the results suggest that father personality plays a role in the development of kitten personality.

Mertens and Turner (1988) confirmed that individual personality types in adult cats influenced their behaviour when interacting with strange people in an unfamiliar setting (an observation room), a nice demonstration of differing "styles" of cat-human interaction. Lowe and Bradshaw (2002) extended these findings by examining how pet cats reacted to being held and petted for one minute by an unfamiliar person, at ages ranging from weaning to three years old. They found stable individual differences in attempts made to escape and others signs of distress (growls, tenseness, protruding claws) from 4 months on. Early handling did not seem to be a major factor after 4 months, suggesting that the way a cat reacts to being handled by an unfamiliar person is largely determined by then. The most escape attempts and distress occurred at 12 months, a time when social and hormonal changes might also be occurring (e.g. Bernstein & Strack 1996).

Mendl and Harcourt (2000) revisited the issue of individuality and socialization, and suggested an overall schema summarizing the factors that appear to affect the expression of "friendliness to humans". These include early social experience with mother and siblings, paternity, breed, coat colour, maternal care, duration and quality of interaction with humans (and probably timing and context as well), and environmental complexity, affecting an individual cat's boldness in novel situations and specific responses to humans. All of these factors together interact with features of a specific current situation to predict the level of "friendliness" that will result. Evidence they present suggests that some behaviours will be stable within an individual over time and some may change.

Siegford *et al.* (2003) have attempted to use the understanding gained from these and other studies (e.g. Lee *et al.* 1983) to develop a relatively quick, simple, and reliable cat temperament test that would help shelter staff, veterinarians, and others to better assess "cat sociability, aggressiveness, and adaptability"(see section 5.3 for more detail).

3.2 Breed Differences

There are currently 37 pedigree and four "miscellaneous" breeds recognized by the Cat Fanciers' Association, Inc. in the United States, described as the world's largest registry of pedigreed cats (see website www.cfainc.org); there are similar numbers listed by the Governing Council of the Cat Fancy in the United Kingdom (see website ourworld.compuserve.com/homepages/GCCF_CATS). Although both groups and many popular cat books describe physical and behavioural characteristics of different breeds, for example the "intelligence, inquisitive personality, and loving nature" of the Siamese or the "playful but not demanding and tremendously responsive nature" of Persians, few studies exist that directly examine behaviour differences among breeds.

In one such study, Turner (2000a) sought to examine breed differences in human-cat interactions by looking specifically at pedigree Siamese and Persian versus non-pedigree cats. The common belief among most cat owners is that these two breeds, among the oldest purebred lines, behave differently from the "common" domestic short hair and are among the most extreme of cat breeds in behaviour and character. Owners of Siamese, Persian, and non-pedigree cats were asked to assess their cats' behaviour traits using a series of rating scales (for example, for playfulness and affection toward owner). They were asked to rate their actual cat versus their "ideal" cat, that is, how their current cat actually behaves versus what behaviour they would most like to see in a cat. They were also directly observed interacting with their cats at home (both techniques were used in part to control for differences in owner perceptions and behaviour). A total of 21 Siamese, 35 Persian, and 61 non-pedigree cat households was observed.

This study confirmed that selective breeding seems to have resulted in Siamese and Persians being more predictable and socially interactive than non-pedigree cats. Siamese initiated more interactions and vocalized more in doing so (confirming what cat most owners believe about this "talkative" breed). Their owners rated them as more playful, curious, friendly to strangers, more often near the owner, more likely to vocalize, higher on affection to the owner, and significantly less "lazy" than did owners of non-pedigree cats. Persian cats were ranked higher on affection to the owner, staying closer to them, vocalizing more, more predictable, clean, fussy about eating, and more friendly toward strangers. Overall, the purebreds were fussier eaters but were ranked as being better behaved and more interested in their owners than non-pedigree cats. Direct observations in the home showed that people tended to interact much longer with the purebreds than with the non-pedigree cats, and spent more time near them. Results also showed that owners tended to interact more often and for longer periods and spoke more frequently with cats who were confined indoors, usually the purebreds, than with cats who went outdoors. Further, older adult humans (greater than 65 years) seemed to accept the independence of their cats better than did younger adults, and seemed more tolerant of them.

Studies such as these provide prospective owners with research-based information about breed behaviour and the effects of genes and early handling on subsequent behaviour. Such information may ultimately guide people in choosing their pet cat more wisely in the future or in dealing with cat behaviour problems.

3.3   Interactions between Cats and Humans

Traditionally, the study of interspecific interactions by animal behaviour researchers has focused on examples such as oxpeckers and oxen, flowers and their pollinators, ants and acacia trees. And yet some of the most obvious interspecific interactions, those between pet owners and their pets, have been largely ignored. Cat-human interactions are no exception; few studies have actually examined how cats interact with one another and with humans in the home. Cat interaction studies by Bernstein and Strack (1996) and by Barry and Crowell-Davis (1999) are two pioneering studies in this area. Both demonstrate the complexity, subtlety, and adaptability of cats that are placed together by humans in groupings and in environments usually not of their own choosing. The role of individual cat personalities and changing circumstances in ensuing behaviour is clear in both studies.

Bernstein and Strack (1996) focused, in over 300 hours of observation, on the spacing patterns, use of favoured spots, dominance and other relationships, and communication behaviours, such as tail signalling, of 14 cats in one home. Individual personalities, age and gender played major roles in the cats' behaviours. For example, the death of the oldest male, the only cat in the group to demonstrate classic dominance behaviours of fighting, chasing others, and supplanting, led to a number of unexpected changes in the spacing patterns and behaviours of other cats, including those with whom this male had not obviously interacted (they had rarely if ever been the targets of his dominance behaviours). Favoured spots throughout the house, where certain cats were likely to be found at certain times of day, were time-shared by specific groups of individuals rather than being used randomly. These groups seemed to be based on gender and individual identity, as well as on past history (as when an older female continued to share a spot with an adult male who she had "adopted" as a kitten). An increasing understanding of such normal behaviours among cats in the home would improve our ability to tell owners what to expect from their cats, how to deal with problems and the role owners play in cat-cat as well as cat-human interactions.

Barry and Crowell-Davis (1999) examined the behaviour of two-cat dyads in 60 homes. Twenty dyads of each gender combination (male-male, male-female, female-female) were observed for 10 hours for each pair, a total of 600 hours. All the cats were neutered and indoor-only. They found less aggression and more affiliative behaviours and time spent in proximity than they had expected. Only 68 instances of aggression were noted, and these seemed to depend more on relationships between specific individuals than on gender, age, size of home or other factors. Length of time in the relationship correlated with decreased levels of aggression. Cats spent an average of 40% of their time within 3 meters of each other, despite their homes having more than enough space to allow them to stay further apart, and male-male pairs spent more time in close proximity, 0 to 1 meter, than did other pairs. A wide variety of affiliative behaviours were seen, including allorubbing, allogrooming, and sniffing. Despite the limited spatial range available and the forced proximity of these cats to each other, they displayed behaviours more typical of a social than of an asocial species. Cat social behaviour is discussed in greater detail in Chapter 1.

Few studies have directly examined interactions between cats and humans in the home. In one such pioneering study, Mertens (1991) showed the complexity of the human-cat relationship as it occurs naturally in the house. She observed 72 cats interacting with 162 people over a 12-month period, in sessions lasting 210 minutes each. She attempted to reduce observer effect by trying to make the owners and cats feel as if she were a normal visitor. She talked to and ate with owners if invited, and stood or sat in rooms as a visitor might; however, she never interacted with or responded to the cats. She identified and examined a list of "social events" engaged in by people and their cats, and the durations of each. The list included a range of typical situations, such as approaching within one meter or withdrawing that far by a cat or human, passing by within sufficient proximity for an interaction, a person picking up or putting down a cat or a cat jumping up on a person's lap or leaving it, petting of a cat by a human and rubbing of a human by a cat, a human speaking to the cat, and so forth. She examined the results by person, by cat, and by the cat-human dyad.

Generally, there was a low level of interaction and most interactions were of fairly short duration (one minute or less). Humans tended to approach within 1 meter of the cat more often than the reverse, but when the cat did the approaching, the human and cat stayed within 1 meter of each other for a longer time. The gender and ages of the people in the experimental situation affected cat interactions: women, men, boys, and girls interacted differently with cats (for example, adults vocalized toward the cat earlier in an interaction and for longer than did children). Women spent more time at home and therefore had more interaction with cats than did men; juvenile humans (11 to 15 years of age) were least likely to be within 1 meter of the cat and had the least amount of interaction, although it was not clear why. Single cats stayed closer to owners for longer and had more social play and more interactions in general with owners than did multiple cats. Interaction, proximity, and rubbing by the cat were moderately more frequent in smaller than in larger families. In view of the increasing popularity of cats as companion animals in many countries, follow-up studies are needed in this area.

Turner also investigated how humans interact with their pets, focusing more on emotional aspects. In his study of purebred versus non-pedigree cats (2000a, see section 3.2), he found that humans tended to have certain expectations of their cats, and that purebreds seemed to meet them more predictably than did non-pedigree cats. Turner's group also demonstrated that the human-cat relationship is indeed a "two-way partnership, with both parties adjusting their behaviour to that of their partners" (Turner & Stammbach-Geering 1990; Turner 2000b).

Heidenberger's survey of 550 German cat owners (1997) also provided insight into how cats and humans interact and utilize the space and resources inside houses (e.g. how much space cats utilize, who sleeps with whom, availability to cats of food and litter boxes). Some information about what owners perceived as problems (behaviour the owner dislikes and wants to change) was also examined and attempts made to find correlations. For example, neutered females (who represented the largest group of cats in the study) were perceived most often as exhibiting problem behaviour, and people who said they interacted with their pets for several hours over the course of each day mentioned fewer problems. However, the author points out that problem behaviour was owner-defined (for example, one owner's problem, such as the cat scratching on furniture, might be seen by another owner simply as typical cat behaviour and not reported), and owners who mentioned fewer problems might have different perceptions of what constitutes acceptable behaviour in cats.

Studies by Alger and Alger (1999, 2003) of cats in shelters echo findings in the home, with both cats and people (mainly the shelter volunteers) making choices about who they will interact with, where, how and when. These authors talk about the "negotiated order" that emerged in the culture of one shelter they studied closely, an order based on both cat and human needs and behaviours. They stressed that the cats were partners in the interactions with each other and with humans, rather than targets.

One of the most obvious and familiar human-cat interactions is that which occurs when humans pet or stroke cats (Figure 2). Surprisingly, little formal study has focused on this interaction. Turner and his colleagues (1999; Turner & Rieger 2001; Turner et al. 2003) have been interested in the importance of petting in providing emotional support for humans, that is whether petting cats can elevate the mood of humans. Results indicate that petting of cats seems to have little effect on mood overall and does not seem to predictably put owners into good moods as petting of dogs seems to, but does seem to help decrease negative mood (see section 2.2).

Figure 2. Surprisingly, little formal study has focused on the interaction which occurs when humans pet or stroke cats.

In a preliminary study based on responses of cat owners to questions about petting, Bernstein (2000) has found patterns of interaction that suggest shared ritual, that is, agreed-upon rules for social interaction and shared routines (see Alger & Alger 1997). Owners identified body areas where their cats seemed to prefer to be petted; cats indicated these areas by engaging in behaviours such as staying still, or closing their eyes, or moving their head or body in such a way as to encourage rubbing of specific sites (for example, along the cheeks, between ears and eyes, top of head, stomach). Owners described sequences of petting that the cat seemed to indicate were desired (e.g. head to back to tail to stomach), and behaviour patterns of the cat that were interpreted as seeking initiation of petting (for example, leap on lap, rub on person's leg, flop down in front of person and look up). Some cats had specific "petting arenas" in the house, often leading owners to a particular spot before standing still or flopping down for petting, or only allowing petting in certain sites, such as the bathroom.

In a different approach, Soennichsen and Chamove (2002) tested nine cats for body-area petting preferences using a prescribed protocol. Family members were instructed to pet their cats for five minutes at one of four body sites, three involving gland areas (temporal gland between the eye and ear, perioral gland on chin and lips, and caudal gland) and one non-gland site. The temporal area had the greatest number of positive responses from the cats (although it is not clear what that response was) and the caudal the least. These results are similar to those in Bernstein's survey, where 48% of 90 cats preferred petting in the head area: 27% of those on the cheeks, nose and eye area or under the chin, and 21% on the ears, in front of the ears (temporal area) or behind the ears; only about 8% of the cats were described as preferring to be petted on the stomach or tail (either at the base of the tail or on the tail itself).

Cat vocalizations toward humans are another obvious area for study. Attempts have been made to parse the vocalizations by context (Moelk 1944, 1979; Brown *et al.* 1978; Bradshaw & Cameron-Beaumont 2000; Nicastro & Owren 2003), and to even assess human perceptions of the calls. For example, Nicastro and Owren (2003) asked human subjects to classify cat vocalizations into specific contexts; accuracy was just above chance, and better for those who had lived with, interacted with, and had a general affinity for cats. This suggests that cat meows are relatively nonspecific and their meanings must be learned by human owners and accompanied by other contextual information to be meaningful. Recently, Nicastro (2004) asked human subjects to rate vocalizations from specific contexts as pleasant or not; long, low-frequency calls were perceived as less pleasant than shorter, higher-frequency calls, indicating an ability of humans to gain important emotional information from the calls, perhaps helping to prompt specific and useful responses. The latter is an attempt, in part, to assess whether domestic cat vocalizations have been adaptively selected to elicit positive responses from humans.

An overview of cat to cat communication by Bradshaw and Cameron-Beaumont (2000) examines a variety of other signals, including olfactory, visual, auditory and tactile displays, as well as vocalizations. These authors also examine the evolutionary relationships among the Felidae and seek patterns in the evolution of various communication signals. They suggest that some signals, such as vocal meows and purrs, visual tail-up signals (previously examined by Kiley-Worthington 1976 and Bernstein & Strack 1996), rolling and certain rubbing displays, have been modified for interspecific communication with humans, including neotenization of some of the signals; these possibilities have only begun to be evaluated.

There is a large veterinary literature on food products for pet cats. Research is ongoing to explore what foods best help cats develop good health, maintain it, and avoid potentially fatal problems (e.g. National Research Council 2003, see Chapter 9). However, little study has been done of the actual feeding interaction itself (initiation, coordination, ending), even though this is one of the most common, frequent and important interspecific interactions in which humans and cats engage and where communication and manipulation by one or both parties may play important roles. Bradshaw and Cook (1996) did examine the behaviour of 36 cats during feeding for a general overview of cat behaviour in this situation. They observed each cat during a pre-eating period as each owner opened a can of food provided by the observer (to begin an observation period), as the cat ate the food, and during the first 5 minutes post-eating. Each cat was observed 8 times, for a total of 288 behaviour sequences. Overall, cats spent much of the time during the pre-meal period interacting with the owner, using communicative patterns such as meowing, tail-up and rubbing, and spent much of the post-meal time grooming, with much less interaction with the owner. There were elements of individual "style" within these sequences, but no major differences among cats. While these behaviour patterns are quite well known to cat owners, this is one of the few studies to seriously research this situation and to document the behaviour quantitatively.

Other common human-cat interactions that have not yet been well studied include those involving litter boxes, such as placement, litter type, number of boxes per social group (there is some evidence that cats try to solicit changing of soiled litter, S. Crowell-Davis, personal communication); contact-seeking behaviour, with either party initiating this (e.g. petting, cats sleeping on laps or on beds with owners); and conversation. While study of the latter has begun for dogs (e.g. Mitchell & Edmonson 1999, Mitchell 2001), there are few data for cats. In a preliminary study, Sims and Chin (2002) observed undergraduate students as they used a toy to engage a cat that was unfamiliar to them. Almost all the students spoke to the cat, and the language used was similar to child-directed language in a number of ways, for example involving short utterances, very short words, a large amount of repetition, and many imperatives. The authors concluded that this interspecific conversation follows a human model whereby the human perceives a social interaction and thus uses speech, modifying that speech based on the perceived comprehension of the listener. It is not clear, however, if or how the cat modifies its own "conversational" style. Further experimental studies, as well as studies of naturally-occurring conversations between cats and humans in the home setting, are needed.

Cats may suffer when their human companions leave them alone in the house. There is some evidence that they develop clinical signs of separation anxiety, a phenomenon more typically associated with dogs, including inappropriate urination and defaecation, excessive vocalization, destructiveness, and other problem behaviours (Schwartz 2002).

Another area that has not been well studied involves interactions that occur between cats and other animals, where the animals interact as if they were companions, for example cats and dogs in the same household that regularly play and sleep together, or cats that regularly stay and interact with horses: are these cases of pets keeping their own pets, of companions having their own companions? While there is much anecdotal information about this phenomenon, research has not directly addressed this aspect of cat relationships.

4. RESPONSIBILITIES OF PET OWNERSHIP

4.1 Veterinary Health Care

There are obvious responsibilities of pet ownership: providing adequate and appropriate nutrition, means of identification to avoid loss, vaccinations, and other preventative health care. Cat owners, however, traditionally have been less likely than dog owners to take their pets to veterinarians. The latest survey conducted by the American Veterinary Medical Association in the United States (AVMA 2002), based on information from 54,000 households in 2001, found that even though cat owners made an astounding 70 million visits to veterinarians, dog owners made over 117 million visits, despite more cats being kept as pets. Nevertheless, according to surveys conducted by the American Pet Products Manufacturers Association (APPMA 2004), cat owners increased the number of visits to the vet from 1.6 in 2000 to 2.3 visits in 2002, making them more comparable to dog owners (who averaged 2.6 visits in 2000 and 2.7 in 2002). In the American Veterinary Medical Association Survey (2002), both dogs and cats were most frequently brought in for physical exams (69% and 67% of visits, respectively). Dog visits were more likely to involve drugs and medications (31% of visits versus 18% for cats), while cat visits involved vaccinations slightly more frequently (71%

for cats versus 64% for dogs). Cat visits were much more likely to involve spaying or neutering (14% versus 6%). Few dogs (0.6%) and cats (0.3%) were provided with computer microchips or tattoos for identification. In the United States in 2001, cat owners spent over 6.6 billion dollars and dog owners over 11.6 billion dollars on veterinary visits. It is interesting to note that in 2002 in the United States only about 1% of cat owners and 2% of dog owners, or 3% in total, had health insurance for their pets compared with 15% in England and 57% in Sweden (APPMA 2004).

4.2 Human Health Concerns

The general public, and physicians, veterinarians and healthcare associates considering Animal-Assisted Therapy for their patients, have an interest in and a need to examine potential health risks from companion animals, including cats. Risks include zoonoses, that is diseases naturally transmitted between animals and humans, as well as allergies, asthma, bite injuries and infections, flea and parasite transmission, and other hazards. These risks have been examined from a variety of viewpoints (e.g. Warner 1984; Hoff *et al.* 1999; Morrison 2001; Brodie *et al.* 2002; Linneberg *et al.* 2003; see Chapter 6). Most references emphasize the low risk of disease associated with cat ownership.

Even those humans with compromised or depressed immune systems, including the sick and very young, may benefit from and be able to continue pet ownership with precautions (e.g. Spencer 1992; Angulo *et al.* 1994; Lappin 2000). The introduction of long-lasting rabies vaccines, the finding that toxoplasmosis can be introduced from a variety of sources, not just from cats, and other factors (such as owners keeping cats indoors so they are less likely to be exposed to parasites or to transmissible diseases) seem to have decreased human concerns about cat zoonoses. Some recent studies have even reported surprising results, such as finding that being exposed at home to pet cats and dogs early in life may have a protective effect against allergy (e.g. Ownby *et al.* 2002) or even asthma (Perzanowski *et al.* 2002), rather than exacerbating these conditions, although one study found pet ownership had little effect on asthma either way (Ownby *et al.* 2002).

Pets have also become one of many foci in the "war on terrorism". There is concern that bioterrorism might include the use of pets and other animals as transmitters of disease. More positively, pets might also serve as a "first alert" repository of diseases, indicating that bioterrorism is being attempted.

5. FAILURES OF THE HUMAN-CAT RELATIONSHIP

5.1 Animal Abusers and Hoarders

Sometimes the human-animal bond goes wrong in serious ways. Animals are often subjected to physical and emotional abuse and cruelty, and they may also become targets of hoarding behaviour, in which individuals live with dozens to hundreds of living and dead animals. Such individuals often show signs of pathological self-neglect and a variety of psychological conditions, including obsessive-compulsive disorder (Arluke *et al.* 2002). They may surround themselves with other objects as well as animals, such as newspapers, dirty plates and utensils, or accumulated food, and declare that they love their pets while seriously neglecting them.

Research presented in Arluke and Luke (1997), Lockwood and Ascione (1998), Ascione and Arkow (1999), Donley *et al.* (1999), and a classic paper by DeViney *et al.* (1983), most vividly examine the connection between animal abuse and violence among humans. Summaries of case reports of abuse, chilling first-person accounts, reviews of laws and the handling of cases by authorities provide important overviews of these phenomena and set an agenda for future work. Although one of the major points of these studies is that there is a connection between abuse of animals and abuse of people, one of the more disconcerting findings is that many instances of animal abuse are not done by individuals who go on to hurt humans. Rather, abusive acts can be performed by "apparently normal" individuals who do not see anything wrong with harming dogs and cats, and who do not see them as deserving of care or respect.

Serpell (1999) examines this phenomenon from a historical perspective. He notes that hunter-gatherers exerted little control over their wild prey, rarely meeting them or directly interacting with them until the moment of death, seeing them as independent beings with independent minds; the need to meet them on their own terms in order to gain food led to a respect for animals. Farmers and herdsmen, on the other hand, have a great deal of control over their domestic animals from the start, a situation which fosters domination as a principle force rather than equality. This human-centered worldview has been reinforced by philosophers since Aristotle, including Aquinas and Descartes. Serpell argues, then, that humaneness, kind or respectful treatment of animals, is an ancient human phenomenon rather than a recent one, and that this natural tendency was suppressed by the rise of domestication and other concurrent cultural changes. Rather than a tendency to abuse animals being part of our "animalistic" nature, this approach suggests the reverse, that we are much more likely naturally to respect animals and therefore can change cultural historical habits that emphasize domination and control.

Several researchers have examined the problems animal abuse pose for veterinarians and suggested strategies and policies for dealing with them (e.g. Ascione & Arkow 1999; Donley *et al.* 1999). Sharpe and Wittum (1999) provide a brief overview of the difficulties veterinarians face in dealing with these issues in small animal practice, issues similar to those faced by medical practitioners confronted with child abuse but without the supporting legal network. Munroe and Thrusfield (2001) presented some quantitative information by surveying 1000 small animal practitioners in the United Kingdom: of 404 responses, 48% had suspected or seen cases of non-accidental injury, with 448 documented cases (243 dogs and 182 cats). The Royal College of Veterinary Surgeons in the United Kingdom has issued guidelines to its practitioners, and the American Veterinary Medical Association issues periodic news reports and educational notes on these issues and includes sessions on this topic at its annual conferences (see website at www.avma.org/onlnews).

Animal collecting or "hoarding" has also been receiving more attention as a major source of animal abuse and an indicator of human mental illness (e.g. Lockwood 1994; Patronek 1999). Hoarders accumulate large numbers of animals, which overwhelm the ability of the person to care for them. The persons involved often fail to acknowledge the deteriorating condition of the animals and the household environment, and fail to see the negative effects on their own health or on that of their family or other housemates. Serpell (2002) notes that the same anthropomorphism that may enable people to see companion animals in ways that facilitate the normal relationship may also play a role in animal abuse and hoarding. When the human-like expectations are not met or when people over-empathize with the animals, abuse and hoarding may result.

According to information provided by the American Veterinary Medical Association, (see website, based on information from Patronek 1999 and Arluke *et al.* 2002), 76% of hoarders are female, nearly half are 60 years of age or older, and most are unmarried and live alone. Animal urine and faeces as well as dead and sick animals are commonly found throughout the house. Although hoarders often accumulate multiple species, cats are involved in 65% of cases. In 1997, an interdisciplinary group including researchers, veterinarians, social workers, and humane society leaders formed the Hoarding of Animals Research Consortium, to develop more effective interventions through research, veterinary support and public education tasks (see website at www.tufts.edu/vet/cfa/hoarding/index.html).

Animal law is a rapidly developing field with a burgeoning number of courses, books, and conferences on the topic. For example, animal rights legislation and law enforcement considerations are included in Ascione and Arkow (1999), with specific recommendations for improvement at all levels of the legal system. Other examples include Francione (1995, 2000); Center for Animals and Public Policy at Tufts University School of Veterinary Medicine, (www.tufts.edu/vet/cfa/legislat.html); the Animal Rights Law Project, Rutgers University, the first law school in the United States to include animal rights in the curriculum ten years ago (www.animal-law.org), the National Center for Animal Law, Lewis and Clark Law School, Portland, Oregon (www.lclark.edu/org/ncal), and its Animal Law Review, the first law review dedicated to this topic; the Animal Law Center, Boulder, Colorado, a nonprofit organization dedicated to ensuring legal rights of animals; and numerous other groups and organizations such as the Animal Legal Defense Fund (www.aldf.org). In 2004 an International Animal Law Conference was held at the California Western School of Law, San Diego, California, United States, featuring scholars and legal authorities from a wide range of fields and a number of countries.

Fortunately, there are increasing numbers of books, research articles, and educational literature and programs designed to examine and change attitudes and increase the humane treatment of animals (e.g. see summaries of programs in Ascione & Arkow 1999 and Fine 2000a; books such as those by Beck & Katcher 1996; Serpell 1996; Dolins 1999; articles such as Ascione & Source 1997; online information such as that of the Scottish Society for the Prevention of Cruelty to Animals at www.scottishspca.org/campaign/firststrike.html and online brochures such as that of the National Society for the Prevention of Cruelty to Children in the United Kingdom, working with 13 affiliated organizations).

5.2 Behaviour Problems

The last decade has seen explosive growth in interest and research in pet behaviour problems (see Chapter 4). Veterinary clinicians, researchers in ethology and psychology, and a variety of other individuals with animal experience (e.g. trainers, breeders) have developed private practices dedicated to helping people deal with pet problems. Specializations in behaviour are now recognized in veterinary medicine, and boards of certification have been formed to certify practitioners from various fields of study (for example, see the American Veterinary Society of Animal Behavior website at www.avma.org/avsab, the Animal Behavior Society website at www.animalbehavior.org, and the website of the Association of Pet Dog Trainers at www.apdt.com).

5.3 Shelter Issues

Interest in pet behaviour has been generated in part by the increasing number of pets, numbering in the millions, that are surrendered to shelters for euthanasia each year because of behaviour problems. Inappropriate elimination (toileting) leads the list of such problems for cats, followed by aggression and destructiveness (e.g. furniture scratching). These problems are themselves often the result of pets being in stressful social situations in homes, the result of crowding, incompatibility among individuals, or lack of owner attention or knowledge (Patronek *et al.* 1996; Scarlett *et al.* 2002).

A study by Patronek *et al.* (1996) helped demonstrate that risk factors for relinquishment are often ones that can be modified with proper intervention and education. Surveys of 218 owners who had relinquished cats to a shelter, versus 459 who had not, revealed several factors that increased the risk of relinquishment. These included the cat remaining sexually intact (possibly from the resultant spraying, yowling, and fighting with other cats, although this was not stated), cats being allowed outdoors, cats being mixed breed rather than purebred, the owner being uneducated about cats (those that read a book or other educational material about cats were less likely to relinquish their cat), and the owner having specific expectations about the cat's role in the household (such as being a close companion). Miller *et al.* (1996) obtained similar results in a smaller study, including the finding that young cats were more likely to be relinquished (for scratching furniture or "aggressive" play), and that an owner's lack of understanding or knowledge of normal feline behaviour often led to unrealistic expectations. The Miller study also noted that when owners had to move unexpectedly to a new location there were often problems, such as housing rental policies that excluded pets, and that these unpredictable aspects also played important roles in relinquishment.

A comprehensive study by Neidhart and Boyd (2002) examined several important adoption issues. They found that retention of adopted animals is similar whether pets were acquired directly from the shelter or at alternative sites, such as large chain pet stores, which is reassuring for shelters attempting to increase adoption rates. But these researchers, like others before them (e.g. Patronek *et al.* 1996), had problems obtaining information on adoption outcomes, despite using a professional marketing firm to make contacts. Only approximately 20% of adopters could be located and were willing to participate in the one-year study. These individuals, however, provided important information about the successes and failures of adoptions. How the pet related to the family, based primarily on pet characteristics and personality, as well as presence and age of children in the home, seemed to best predict retention. Factors for relinquishment, disappointment with or loss of the animal during the year differed for dogs and cats. Behaviour problems were more likely to be cited for dogs than for cats. Overall, more adopters were satisfied with their new cat than dog adopters were with their dogs, although satisfaction was high for both groups (94% for cats, 86% for dogs). Unfortunately, cats died during the first year of adoption at twice the rate of dogs, presumably from collisions with cars (although only a small number of animals, eight dogs and 15 cats, actually perished). Findings from these and other studies provide the impetus for veterinarians and others to provide guidance, support and education to owners in an effort to modify owner perceptions and behaviour.

Since the 1980s, there has been a movement towards no-kill shelters, perhaps best illustrated by the pioneering work of the San Francisco Society for the Prevention of Cruelty to Animals (San Francisco SPCA, website at www.sfspca.org). Once accepted into the shelter, animals are not euthanized unless they are incurably sick, disabled or display extreme behaviour that makes them unsuitable for adoption. However, due to limited space, these facilities are often quite selective in which animals they take. Animal sanctuaries specialize in offering lifetime care and are also selective. Most shelters take in as many animals as possible as a community service, and must then choose which to keep and which must be euthanized. The Humane Society of the United States and the American Society for the Prevention of Cruelty to Animals play increasingly important roles as educators about animal care and animal welfare issues and as sponsors of studies of behaviour, in an attempt to decrease the numbers of animals entering shelters and increase the number of animals adopted from them. A growing movement seeks to couple no-kill policies with attempts to better observe, evaluate and modify the behaviour of shelter animals so they will be more adoptable and less likely to be returned.

In an effort to help shelters, veterinarians, and others who need to assess cat temperament, Siegford *et al.* (2003) developed and validated a strategy for testing cats. Working with 10-month old kittens housed in an animal facility, the researchers tested behaviour over the next 8 months, both before adoption and for 3 and 6 months after adoption. Individuals were evaluated in three ways before adoption, using an easily-scored Feline Temperament Profile (FTP) developed by Lee *et al.* (1983), video tapes of cat interactions with their caretakers at the animal facility, and reactions of cats to an unfamiliar man and woman in open field tests in an unfamiliar room. The FTP was then administered again post-adoption. The researchers found that cat scores on the FTP were fairly consistent over time and circumstance and correlated positively with responses to caretakers and unfamiliar humans. Cats could be ranked generally by their FTP scores as being more or less sociable toward people. This provided insight into what kinds of reactions shelter workers, veterinarians, and new owners might expect from individual cats, and what steps they might take to avoid problems or incompatibilities. For example, cats that ranked as less sociable on all tests might best be placed by a shelter with an experienced cat owner or someone who did not expect or desire a social, attention-seeking companion; a veterinarian might handle such a cat differently when it came in for a check-up.

The cat overpopulation problem, relinquishment of cats to shelters and factors affecting the adoption of cats from shelters are discussed in greater detail in Chapter 5.

5.4 Allowing Cats Outdoors

Traditionally in the United States cats have been allowed to roam free at their will (Figure 3). In the last decade or so, this practice has changed in response to a number of pressures. Research suggesting that cats are incredibly successful predators with the ability to reduce wild bird populations considerably, at least on islands (e.g. see Fitzgerald 1988, but see tempered overview of cat predation, Fitzgerald & Turner 2000) has resulted in the call by many humane organizations, wildlife conservation groups, and ornithological associations to ban cats from the outdoors. Increasing risks to cats from their own predators, such as raptors and coyotes, and the growing volume of vehicular traffic have also affected owner willingness to allow cats to roam. An increase in infectious and often lethal diseases, such as feline immune deficiency syndrome, feline leukaemia, and infectious peritonitis, has resulted in people wanting to protect their cats from infected strays. Owner fears that they themselves might be susceptible to cat diseases also plays a role in keeping cats indoors, although there is little evidence for significant zoonotic transmission (see section 4.2 and Chapter 8).

In one moderately sized American city, Akron, Ohio, a series of letters to the editor in response to the question "should cats be required to be licensed and leashed" were intensely anti-cat and in favour of restricting their freedom. Most letter writers complained about cat faeces in their gardens, cats digging in flower beds, or simply that cats were wandering around and were considered scary, annoying or possible carriers of disease (Akron Beacon Journal 2001).

Figure 3. In the United States, cats have traditionally been allowed to roam free at their will, but in the last decade this practice has changed; more and more cats are being confined indoors. (Courtesy of Cerian Webb).

This combination of issues and concerns seems to be having an impact on owners, at least in the United States. In a preliminary study of 256 households surveyed from 1993 to 2003, Bernstein (2001, 2003) found that 50% of 503 cats were being kept indoors at all times. Of those allowed outdoors, only 33% were unrestricted (about 17% of all cats); an additional 15% (7% of all cats) had restricted outdoor access, such as sitting with their owners on a house deck, being walked on leashes, being restricted to a lead in the yard, or only allowed into small fenced-in areas in the yard. Interestingly, at least some cats restricted themselves, either showing no interest in going out when offered, or acting fearful and running away when the possibility was presented. Owners of declawed cats were equally likely to let their cats out as keep them in. These findings are in sharp contrast to the figures recently released by the Feline Advisory Bureau (2004, and at www.fabcats.org) from a survey of 1853 British cat owners: in this group, 75% of cats were allowed out at will during daylight hours.

A survey by Clancy *et al.* (2003) of cat owners was conducted in 2001 during routine veterinary visits at the small animal hospital of the Tufts University School of Veterinary Medicine. Based on 184 cats, researchers found that cats acquired recently were less likely to be allowed outdoors than those acquired during previous years. Access to the outdoors, when allowed, was likely to be limited to daytime hours; declawing, age, and health status played no significant role in the decision, and cats acquired as strays were more likely to be allowed outdoors than those acquired at shelters. The latter may be due in part to a growing number of shelters and rescue groups requiring adopters to pledge to keep cats indoors. Perception of cat experience might also play a role, as strays may be perceived as already having demonstrated knowledge of the outdoors and the skills necessary to survive, and therefore be let out more. Both Clancy et al. (2003) and Rochlitz (2003) have illustrated the perils that can befall cats, particularly young males, that go outdoors, including cat bites, disease, predation by hawks and coyotes, and injuries and death from dogs and automobiles. What remains to be tested is whether cats that are kept primarily indoors or are indoor-only are more prone to develop behavioural and other problems. Some evidence suggests this may be the case: in a survey of German cat owners, Heidenberger (1997) found that owners who let their cats out only rarely or irregularly (e.g. only in good weather) were more likely to say their cats had behaviour problems than owners whose cats were allowed out regularly (e.g. whenever they wanted or at least two to three times per week or every weekend; see Chapter 4 on behaviour problems and Chapter 7 on the housing of cats).

5.5 Feral and Stray Cats

Another increasing problem is that of roaming, abandoned, stray, and feral cats (see Chapter 6). In response to growing populations of stray or feral cats, many towns now have their own corps of dedicated "cat caretakers", local people who not only feed the cats and generally attempt to care for them, but also try to catch them and have as many as possible neutered and spayed, often paying veterinary expenses out of pocket.

Major animal welfare groups such as the Humane Society of the United States, the American Society for the Prevention of Cruelty to Animals, and Alley Cat Allies also provide information, education, and guidelines about this growing problem. In 1996 the American Veterinary Medical Association, after holding its Animal Welfare Forum, various hearings, and study of reports from various groups, adopted its first position statement on abandoned and feral cats (AVMA 1996). Although the main goal of the policy is to eliminate the problem, it provides guidelines for management of feral colonies and stresses the need for communities to establish cooperative resource networks (teams of care givers, veterinarians, public health officials, control officers, and others) to work together to achieve the goal of decreasing colony numbers.

6. CONCLUSIONS

Companion animals are important to humans in many ways, bringing a range of health benefits and playing an important role in their lives, particularly within families and with children. While a large array of studies has provided us with insights into the human-animal relationship, fewer of them have focused on cats as opposed to dogs. Nevertheless, while differing from dogs in what they bring to the relationship, cats are clearly very effective companion animals to humans, as evidenced by the increasing numbers of pet cats in many parts of the world. Recent studies have shed light on some of the ways humans and cats interact, and the factors that influence these interactions.

The growing numbers of cats in shelters, and increasing populations of stray and feral cats, are evidence of failures of the human-cat relationship, as is the serious problem of animal abuse and hoarding. These important issues demand that we learn more about our cat companions, so that we can better educate and support the humans who attempt to live with them.

7. REFERENCES

Adamec, R.E., Stark-Adamec, C. and Livingston, K.E. (1983) The expression of an early developmentally emergent defensive bias in the adult domestic cat (*Felis catus*) in non-predatory situations. *Applied Animal Ethology* **10**, 89-108.

Akron Beacon Journal (2001) Should cats be licensed, collared and otherwise restricted? Wednesday Forum, June 13.

Albert, A. and Anderson, M. (1997) Dogs, cats, and morale maintenance: some preliminary data. *Anthrozoös* **X**, 121-124.

Alger, J.M. and Alger, S.F. (1997) Beyond Mead: symbolic interaction between humans and felines. *Society & Animals* **5**, 65-81.

Alger, J.M. and Alger, S.F. (1999) Cat culture, human culture: an ethnographic study of a cat shelter. *Society & Animals* **7**, 199-219.

Alger, J.M. and Alger, S.F. (2003) *Cat Culture: The Social World of a Cat Shelter*. Temple University Press, Philadelphia, USA.

American Pet Products Manufacturers Assocation (2004) *National Pet Owners Survey*. Greenwich, CT, USA, xxiv.

American Veterinary Medical Association (1996) AVMA adopts position on abandoned and feral cats: AVMA position statement on abandoned and feral cats. *J. American Veterinary Medical Association*, **209**, 1042-1043.

American Veterinary Medical Associaton (2002) *U.S. Pet Ownership & Demographics Sourcebook*. Membership and Field Services, Schaumberg, IL, USA.

Angulo, F.J., Glaser, C.A., Juranek, D.D., Lappin, M.R., and Regnery, R.L. (1994) Caring for pets of immunocompromised persons. *J. American Veterinary Medical Association* **205**, 1711-1718.

Archer, J. (1997) Why do people love their pets? *Ethology and Sociobiology* **18**, 237-259.

Arluke, A. (1997) Links between animal and human abuse. *Abstracts of the International Society for Anthrozoölogy*, 6th Annual Conference, Boston, MA, USA..

Arluke, A. and Luke, C. (1997) Physical cruelty toward animals in Massachusetts, 1975-1996. *Society & Animals* **5**, 195-204.

Arluke, A., Frost, R., Luke, C., Messner, E., Nathanson, J., Patronek, G.J., Papazian, M. and Steketee, G. (2002). Health implications of animal hoarding. *Health & Social Work* **27**, 125-136.

Ascione, F. R. and Arkow, P. (1999) *Child Abuse, Domestic Violence, and Animal Abuse.* Purdue University Press, West Lafayette, Indiana.

Ascione, F. R., Kaufmann, M.E. and Brooks, S.M (2000). Animal abuse and developmental psychopathology: recent research, programmatic, and therapeutic issues and challenges for the future. In Fine, A. (ed.). *Handbook On Animal-Assisted Therapy: Theoretical Foundations and Guidelines for Practice,* Academic Press, San Diego, USA, pp. 325-354.

Ascione, F.R. and Source, R (1997). Humane education research: evaluating efforts to encourage children's kindness and caring toward animals. *Genetic, Social & General Psychology Monographs* **123**, 59-78.

Ash, S.J. and Adams, C.E. (2003) Public preferences for free-ranging domestic cat (Felis catus) management options. *Wildlife Society Bulletin* **31**, 334-339.

Bahlig-Pieren, Z. and Turner, D.C. (1999) Anthropomorphic interpretations and ethological descriptions of dog and cat behaviour by lay people. *Anthrozoös* **12**, 205-210.

Barry, K. and Crowell-Davis, S. (1999) Gender differences in the social behaviour of the neutered indoor-only domestic cat. *Applied Animal Behaviour Science* **64**, 193-211.

Baun, M. M. and McCabe, B.W. (2000). The role animals play in enhancing quality of life for the elderly. In Fine, A. (ed.), *Handbook On Animal-Assisted Therapy: Theoretical Foundations and Guidelines for Practice,* Academic Press, San Diego, USA, pp. 237-251.

Beck, A. (2000). The use of animals to benefit humans: Animal-assisted therapy. In Fine, A. (ed.), *Handbook On Animal-Assisted Therapy: Theoretical Foundations and Guidelines for Practice,* Academic Press, San Diego, USA, pp. 21-40.

Beck, A. and Katcher, A. (1996) *Between pets and people: the importance of animal companionship.* Revised 1st edn. Purdue University Press, West Lafayette, USA.

Bekoff, M., Gruen, L.,Townsend, S.E. and Rollin, B.E. (1992) Animals in science: some areas revisited. *Animal Behaviour* **44**, 473-484.

Bernstein, P.L. (2000) People petting cats: a complex interaction. *Abstracts of the Animal Behaviour Society*, Annual Conference, Atlanta, GA, USA, 9.

Bernstein, P. L. (2001) Cat owners favor keeping cats indoors. *Abstracts of the International Society for Anthrozoology*, 10th Annual Conference, Davis, CA, USA, 10.

Bernstein, P. L. (2003) Cats, houses, and people. *Abstracts of the International Society for Anthrozoology*, 12th Annual Conference, Canton, OH USA, 14.

Bernstein, P.L., Friedmann, E. and Malaspina, A. (2000) Animal-assisted therapy enhances resident social interaction and initiation in long-term care facilities. *Anthrozoös* **13**, 213-224.

Bernstein, P.L. and Strack, M. (1996) A game of cat and house: spatial patterns and behaviour of 14 domestic cats (*Felis catus*) in the home. *Anthrozoös* **IX**, 25-39.

Bonas, S., McNicholas, J. and Collis, G.M. (2000) Pets in the network of family relationships: an empirical study. In Podberscek, A.L., Paul, E.S. and Serpell, J.A. (eds.), *Companion*

Animals and Us: Exploring the Relationships Between People & Pets, Cambridge University Press, Cambridge, pp. 209-236.

Bradshaw, J.W.S. and Cameron-Beaumont, C. (2000) The signaling repertoire of the domestic cat and its undomesticated relatives. In Turner, D.C. and Bateson, P. (eds.). *The Domestic Cat: the biology of its behaviour*, 2nd edn., Cambridge University Press, Cambridge, pp. 68-93.

Bradshaw, J.W.S. and Cook, S.E. (1996) Patterns of pet cat behaviour at feeding occasions. *Applied Animal Behaviour Science* **47**, 61-74.

Brodie, S.J., Billy, F.C. and Shewring, M. (2002) An exploration of the potential risks associated with using pet therapy in healthcare settings. *J. Clinical Nursing* **11**, 444-457.

Brown, K.A., Buchwald, J.S., Johnson, J.R., and Mikolich, D.J. (1978) Vocalization in the cat and kitten. *Developmental Psychobiology* **11**, 559-570.

Carlstead, K., Brown, J.L. and Strawn, W. (1993) Behavioural and physiological correlates of stress in laboratory cats. *Applied Animal Behaviour Science* **38**, 143-158.

Clancy, E.A., Moore, A.S. and Bertone, E.R. (2003) Evaluation of cat and owner characteristics and their relationships to outdoor access of owned cats. *J. American Veterinary Medical Association* **222**, 1541-1545.

Cohen, S. P. (2002) Can pets function as family members? *Western J. Nursing Research* **24**, 621-638.

Collard, R. R. (1967) Fear of strangers and play behaviour in kittens with varied social experience. *Child Development* **38**, 877-891.

Copeland, M.W. (2003) The defining difference: response to "What is a pet?" *Anthrozoös* **16**, 111-113.

Coppinger, R. and Coppinger, L. (1998) Differences in the behaviour of dog breeds. In Grandin, T. (ed.) *Genetics and the Behaviour of Domestic Animals*, Academic Press, San Diego, USA, pp. 167-202.

Coppinger, R. and Schneider, R. (1995) The evolution of working dog behaviour. In Serpell, J. (ed.) *The Domestic Dog: Its Evolution, Behaviour and Interactions with People*, Cambridge University Press, Cambridge, UK, pp. 21-47.

Davis, H., Irwin, P., Richardson, M., and O'Brien-Malone, A. (2003) When a pet dies: religious issues, euthanasia and strategies for coping with bereavement. *Anthrozoös* **16**, 57-74.

DeViney, E., Dickert, J. and Lockwood, R. (1983) The care of pets within child abusing families. *International J. for the Study of Animal Problems* **4**, 321-329.

Dolins, F.L. (ed.) (1999) *Attitudes to Animals: Views in Animal Welfare*. Cambridge University Press, Cambridge.

Donley, L. Patronek, G.J. and Luke, C. (1999) Animal abuse in Massachusetts: a summary of case reports at the MSPCA and attitudes of Massachusetts veterinarians. *J. Applied Animal Welfare Science* **2**, 59-74.

Dresser, N. (2000) The horse bar mitzvah: a celebratory exploration of the human-animal bond. In Podberscek, A.L., Paul, E.S. and Serpell, J.A. (eds.), *Companion Animals and Us: Exploring the Relationships Between People & Pets*, Cambridge University Press, Cambridge, pp. 90-107.

Eddy, T.J. (2003a) What is a pet? *Anthrozoös* **16**, 98-105.

Eddy, T.J. (2003b) The challenge: reflections on responses to "What is a pet?". *Anthrozoös* **16**, 127-134.

Enders-Slegers, M-J. (2000) The meaning of companion animals: qualitative analysis of the life histories of elderly cat and dog owners. In Podberscek, A.L., Paul, E.S. and Serpell, J.A. (eds.), *Companion Animals and Us: Exploring the Relationships Between People & Pets*, Cambridge University Press, Cambridge, pp. 237-256.

Euromonitor International (2003). *The Global Market for Pet Food and Pet Care Products.* London, UK

Feaver, J., Mendl, M. and Bateson, P. (1986) A method for rating the individual distinctiveness of domestic cats. *Animal Behaviour* **34**, 1016-1025.

Feline Advisory Bureau (2004) *Up Close and Purrsonal, Report of the Cat Personality Survey.* FAB Publications, Tisbury, Wiltshire, UK.

Fine, A.H. (2000a). *Handbook On Animal-Assisted Therapy: Theoretical Foundations and Guidelines for Practice.* Academic Press, San Diego, USA.

Fine, A.H. (2000b). Animals and therapists: incorporating animals in outpatient pyschotherapy. In Fine, A. (ed.), *Handbook On Animal-Assisted Therapy: Theoretical Foundations and Guidelines for Practice,* Academic Press, San Diego, USA, pp. 179-211.

Fitzgerald, B.M. (1988) Diet of domestic cats and their impact on prey populations. In Turner, D.C. and Bateson, P. (eds.). *The Domestic Cat: the biology of its behaviour,* 1st edn., Cambridge University Press, Cambridge, pp. 123-147.

Fitzgerald, B.M. and Turner, D.C. (2000) Hunting behaviour of domestic cats and their impact on prey populations. In Turner, D.C. and Bateson, P. (eds.). *The Domestic Cat: the biology of its behaviour,* 2nd edn., Cambridge University Press, Cambridge, pp. 152-175.

Francione, G.L. (1995) *Animals, Property and the Law.* Temple University Press, Philadelphia.

Francione, G. L. (2000) *Introduction to Animal Rights: Your Child or the Dog?* Temple University Press, Philadelphia.

Friedmann, E. (2000) The animal-human bond: health and wellness. In Fine, A. (ed.), *Handbook On Animal-Assisted Therapy: Theoretical Foundations and Guidelines for Practice,* Academic Press, San Diego, USA, pp. 41-58.

Friedmann, E., Thomas, S.A. and Eddy, T.J. (2000) Companion animals and human health: physical and cardiovascular influences. In Podberscek, A.L., Paul, E.S. and Serpell, J.A. (eds.), *Companion Animals and Us: Exploring the Relationships Between People & Pets,* Cambridge University Press, Cambridge, pp. 125-142

Goodwin, D., Bradshaw, J.W.S., and Wickens, S.M. (1997) Paedomorphosis affects agonistic visual signals of domestic dogs. *Animal Behaviour* **53**, 297-304.

Granger, B.P. and Kogan, L. (2000). Animal-assisted therapy in specialized settings. In Fine, A. (ed.), *Handbook On Animal-Assisted Therapy: Theoretical Foundations and Guidelines for Practice,* Academic Press, San Diego, USA, pp. 213-236.

Greenfield, C.L., Johnson, A.L., Schaeffer, D.J., and Hungerford, L.L. (1995) Comparison of surgical skills of veterinary students trained using models or live animals. *J. American Veterinary Medical Association* **206**, 1840-1845.

Hart, L.A. (1990) Pets, veterinarians and clients: communicating the benefits. In *Pets, benefits and practice,* Waltham Symposium 20, BVA Publications, London, pp. 36-43.

Hart, L.A. (1998) *Responsible Conduct with Animals in Research.* Oxford University Press, Oxford.

Hart, L.A. (2000a) Psychosocial benefits of animal companionship. In Fine, A. (ed.), *Handbook On Animal-Assisted Therapy: Theoretical Foundations and Guidelines for Practice,* Academic Press, San Diego, USA, pp. 59-78.

Hart, L.A. (2000b) Methods, standards, guidelines and considerations in selecting animals for animal-assisted therapy. Part A: Understanding animal behaviour, species, and temperament as applied to interactions with specific populations. In Fine, A. (ed.), *Handbook On Animal-Assisted Therapy: Theoretical Foundations and Guidelines for Practice,* Academic Press, San Diego, USA, pp. 81-97.

Hart, L.A. (2003) Pets along a continuum: response to "What is a pet?". *Anthrozoös* **16**, 118-122.

Hart, L.A., Wood, M.W., Massey, A. and Smith, M. (2004) Uses of animals and alternatives in pre-college education in the United States: need for leadership on educational resources and guidelines. *Proceedings of Fourth World Congress, Transforming middle and high school education with animal alternatives, Alternatives to Laboratory Animals* **32** Supplement 1, 485-489.

Hart, L.A. and Wood, M.W. (2004) Uses of animals and alternatives in college and veterinary education at the University of California, Davis: institutional commitment for mainstreaming alternatives. *Proceedings of Fourth World Congress, Transforming middle and high school education with animal alternatives, Alternatives to Laboratory Animals* **32** Supplement 1, 617-620.

Heidenberger, E. (1997) Housing conditions and behavioural problems of indoor cats as assessed by their owners. *Applied Animal Behaviour Science* **52**, 345-364.

Hoff, G.L., Brawley, J. and Johnson, K. (1999) Companion animal issues and the physician. *Southern Medical J.* **92**, 651-660.

Hunt, S.J. and Hart, L.A. (1992) Role of small animals in social interactions between strangers. *The J. Social Psychology* **132**, 245-256.

Johnson, T.P. Garrity, T.F. and Stallones, L. (1992) Psychometric evaluation of the Lexington attachment to pets scale (LAPS). *Anthrozoös* **5**, 160-175.

Karsh, E.B. (1984) The effects of early and late handling on the attachment of cats to people. In Anderson, R.K., Hart, B.L. and Hart, L.A. (eds.) *The Pet Connection, Conference Proceedings*. Globe Press, St. Paul.

Karsh, E.B. and Turner, D.C. (1988). The human-cat relationship. In Turner, D.C. and Bateson, P. (eds.). *The Domestic Cat: the biology of its behaviour*, 1st edn., Cambridge University Press, Cambridge, pp. 159-177.

Katcher, A.H. (1981) Interactions between people and their pets: form and function. In Fogle, B. (ed.), *Interrelationships Between People and Pets*. Charles C. Thomas, Springfield, USA, pp. 41-67.

Katcher, A.H. and Wilkins, G.G. (2000). The centaur's lessons: therapeutic education through care of animals and nature study. In Fine, A. (ed.), *Handbook On Animal-Assisted Therapy: Theoretical Foundations and Guidelines for Practice*, Academic Press, San Diego, USA, pp. 153-177.

Kiley-Worthington, M. (1976) The tail movements of ungulates, canids, and felids with particular reference to their causation and function as displays. *Behaviour* **56**, 69-115.

Lappin, M.R. (2001) Cat ownership by immunosuppressed people. In August, J.R. (ed.), *Consultations in Feline Internal Medicine* 4th edn., W.B. Saunders Company, Philadelphia, USA, pp. 18-27.

Lawrence, E.A. (2003) Some observations on "What is a pet?". *Anthrozoös* **16**, 123-126.

Lee, R.L., Zeglen, M.E., Ryan, T., and Hines, L.M. (1983) Guidelines: animals in nursing homes. *California Veterinarian* **3**: 22a-26a.

Levinson, B. (1969) *Pet-oriented Child Psychotherapy*. Charles C. Thomas, Springfield, IL, USA.

Linneberg, A., Nielsen, N.J., Madsen, F., Frolund, L., Dirksen, A., and Jorgensen, T. (2003) Pets in the home and the development of pet allergy in adulthood. The Copenhagen Allergy Study. *Allergy* **58**, 21-27.

Lockwood, R. (1994) The psychology of animal collectors. *American Animal Hospital Association Trends Magazine* **9**, 18-21.

Lockwood, R. and Ascione, F.R. eds. (1998) *Cruelty to Animals and Interpersonal Violence. Readings in Research and Application*. Purdue University Press, West Lafayette.

Lowe, S.E. and Bradshaw, J.W.S. (2001) Ontogeny of individuality in the domestic cat in the home environment. *Animal Behaviour* **61**, 231-237.

Lowe, S.E. and Bradshaw, J.W.S. (2002) Responses of pet cats to being held by an unfamiliar person, from weaning to three years of age. *Anthrozoös* **15**, 69-79.

Mallon, G.P., Ross, S.B. Jr. and Ross, L. (2000). Designing and implementing animal-assisted therapy programs in health and mental health organizations. In Fine, A. (ed.), *Handbook On Animal-Assisted Therapy: Theoretical Foundations and Guidelines for Practice*, Academic Press, San Diego, USA, pp. 115-127.

McCune, S. (1995) The impact of paternity and early socialization on the development of cats' behaviour to people and novel objects. *Applied Animal Behaviour Science* **45**, 109-124.

McGreevy, P.D. and Nicholas, F.W. (1999) Some practical solutions to welfare problems in dog breeding. *Animal Welfare* **8**, 329-341.

McNally, R.J. and Steketee, G.S. (1985) The etiology and maintenance of severe animal phobias. *Behaviour Research and Therapy* **23**, 431-435.

Meier, G.W. (1961) Infantile handling and development in Siamese kittens. *J. Comparative Physiology and Psychology* **54**, 284-286.

Meier, M. and Turner, D.C. (1985) Reactions of home cats during encounters with a strange person: evidence for two personality types. *J. Delta Society* (later *Anthrozoös*) **2**, 45-53.

Melson, G.F. (2000). Companion animals and the development of children: implications of the biophilia hypothesis. In Fine, A. (ed.), *Handbook On Animal-Assisted Therapy: Theoretical Foundations and Guidelines for Practice*, Academic Press, San Diego, USA, pp. 375-383.

Melson, G.F. (2001). *Why the Wild Things Are: Animals In the Lives of Children*. Harvard University Press, Cambridge, USA.

Mendl, M. and Harcourt, R. (1988) Individuality in the domestic cat. In Turner, D.C. and Bateson, P. (eds.). *The Domestic Cat: the biology of its behaviour*, 1st edn., Cambridge University Press, Cambridge, pp. 159-177.

Mendl, M. and Harcourt, R. (2000) Individuality in the domestic cat: origins, development and stability. In Turner, D.C. and Bateson, P. (eds.). *The Domestic Cat: the biology of its behaviour*, 2nd edn., Cambridge University Press, Cambridge, pp. 47-64.

Mertens, C. (1991) Human-cat interactions in the home setting. *Anthrozoös* **4**, 214-231.

Mertens, C. and Turner, D.C. (1988) Experimental analysis of human-cat interactions during first encounters. *Anthrozoös* **2**, 83-97.

Miller, D.D., Staats, S.R., Partlo, C., and Rada, K. (1996) Factors associated with the decision to surrender a pet to an animal shelter. *J. American Veterinary Medical Association* **209**, 738-742.

Miller, M. and Lago, D. (1990) Observed pet-owner in-home interactions: species differences and association with the pet relationship scale. *Anthrozoös* **4**, 49-54.

Mitchell, R.W. (2001) Americans' talk to dogs: similarities and differences with talk to infants. *Research on Launguage and Social Interaction* **34**, 183-210.

Mitchell, R.W. and Edmonson, E. (1999) Functions of repetitive talk to dogs during play: control, conversation, or planning? *Society & Animals* **7**, 55-81.

Moelk, M. (1944). Vocalizing in the house-cat: A phonetic and functional study. *American J. Psychology* **57**, 184-204.

Moelk, M. (1979) The development of friendly approach behaviour in the cat: A study of kitten-mother relations and the cognitive development of the kitten from birth to eight weeks. In Rosenblatt, J.S., Hinde, R.A., Beer, C. & Busnel, M. (eds.) *Advances in the Study of Behaviour*, vol. 10, Academic Press, New York, USA, pp. 163-224.

Morris, P., Fidler, M. and Costall, A. (2000) Beyond anecdotes: an empirical study of "anthropomorphism". *Society & Animals* **8**, 151-166.

Morrison, G. (2001) Zoonotic infections from pets. *Postgraduate Medicine* **110**, 24-35.

Munroe, H.M.C. and Thrusfield, M.V. (2001) 'Battered pets': features that raise suspicion of non-accidental injury. *J. Small Animal Practice* **42**, 218-226.
Myers, G. (1999) *Children and Animals: Social Development and Our Connection to Other Species.* Westview Press, Boulder, CO, USA.
National Research Council, Subcommittee on dog and cat nutrition (2003) *Nutrient requirements of dogs and cats.* The National Academies Press, Washington, D.C.
Neidhart, L. and Boyd, R. (2002) Companion animal adoption study. *J. Applied Animal Welfare Science* **5**, 175-192.
Nicastro, N. (2004) Adaptive implications of species-level differences in meow vocalizations. *J. Comparative Psychology* **118**, 287-296.
Nicastro, N. and Owren, M.J. (2003) Classification of domestic cat (*Felis catus*) vocalizations by naïve and experienced human listeners. *J. Comparative Psychology* **117**, 44-52.
Ownby, D.R., Johnson, C.C. and Peterson, E.L. (2002) Exposure to dogs and cats in the first year of life and risk of allergic sensitization at 6-7 years of age. *J. American Medical Association* **288**, 963-972.
Parslow, R.A. and Jorm, A.F. (2003) The impact of pet ownership on health and health service use: results from a community sample of Australians aged 40-44 years. *Anthrozoös* **16**, 43-56.
Patronek G. (1999) Hoarding of animals: an underrecognized public health problem in a difficult-to-study population. *Public Health Reports* **114**, 81-87.
Patronek G.J., Glickman, L.T., Beck, A.M., McCabe, G.P. and Ecker, C. (1996) Risk factors for relinquishment of cats to an animal shelter. *J. American Veterinary Medical Association* **209**, 582-588.
Perzanowski, M.S., Ronmark, E., Platts-Mills, T.A.E. and Lundback, B. (2002) Effect of cat and dog ownership on sensitization and development of asthma among preteenage children. *American J. Respiratory and Critical Care Medicine* **166**, 696-702.
Podberscek, A.L. and Gosling, S.D. (2000) Personality research on pets and their owners: conceptual issues and review. In Podberscek, A.L., Paul, E.S. and Serpell, J.A. (eds.), *Companion Animals and Us: Exploring the Relationships Between People & Pets,* Cambridge University Press, Cambridge, pp. 143-167.
Podberscek, A.L., Paul, E.S. and Serpell, J.A. (2000) *Companion Animals and Us: Exploring the Relationships Between People & Pets,* Cambridge University Press, Cambridge.
Rajack, L.S. (1997) Pets and human health: the influence of pets on cardiovascular and other aspects of owners' health. Ph.D. thesis, University of Cambridge, UK.
Rajecki, D.W., Rasmussen, J.L., Sanders, C.R., Modlin, S.J. and Holder, A.M. (1999) Good dog: aspects of humans' causal attributions for a companion animal's social behaviour. *Society & Animals* **7**, 17-35.
Reisner, I.R., Houpt, K.A., Erb, H.N., and Quimby, F.W. (1994) Friendliness to humans and defensive aggression in cats: the influence of handling and paternity. *Physiology & Behaviour* **55**, 1119-1124.
Rieger, G. and Turner, D.C. (1999) How depressive moods affect the behaviour of singly living persons toward their cats. *Anthrozoös* **12**, 224-233.
Ritvo, H. (1985) Animal pleasures: popular zoology in eighteenth and nineteenth century England. *Harvard Library Bulletin* **33**, 239-279.
Rochlitz, I. (2000) Feline welfare issues. In Turner, D.C. and Bateson, P. (eds.). *The Domestic Cat: the biology of its behaviour,* 2nd edn., Cambridge University Press, Cambridge, pp. 208-226.
Rochlitz, I. (2003) Study of factors that may predispose domestic cats to road traffic accidents: Part 1. *Veterinary Record* **153**, 549-553.
Rollin, B.E. and Rollin, L. (2003) Response to "What is a pet?". *Anthrozoös* **16**, 106-110.

Rosenblatt, J.S., Turkewitz, G. and Schneirla, T.C. (1961) Early socialization in the domestic cat as based on feeding and other relationships between female and young. In Foss, B.M. (ed.) *Determinants of Infant Behaviour*, Methuen, London, U.K., pp 51-74.

Salman, M.D., New, J.G., Scarlett, J.M. (1998) Human and animal factors related to the relinquishment of dogs and cats in 12 selected animal shelters in the United States. *J. Applied Animal Welfare Science* **2**, 207-226.

Sanders, C.R. (2003) Whose pet? Comment on Timothy Eddy, "What is a pet?". *Anthrozoös* **16**, 114-117.

Schwartz, S. (2002) Separation anxiety syndrome in cats: 136 cases (1991-2000). *J. American Veterinary Medical Association* **220**, 1028-1033.

Scarlett, J., Salman, M.D., New, J. G., and Kass, P. (2002) The role of veterinary practitioners in reducing dog and cat relinquishments and euthanasias. *J. American Veterinary Medical Association* **220**, 306-311.

Serpell, J. (1985) Best friend or worst enemy: cross-cultural variation in attitudes to the domestic dog. *The Human-Pet Relationship: Proceedings of the International Symposium*, Vienna: Austrian Academy of Sciences/IEMT.

Serpell, J. (1991) Beneficial effects of pet ownership on some aspects of human health and behaviour. *J. Royal Society of Medicine* **84**, 717-720.

Serpell, J. (1996) *In the Company of Animals*, 2nd ed., Cambridge University Press, Cambridge.

Serpell, J. (1999) Working out the beast. An alternative history of western humaneness. In Ascione, F.R. and Arkow, P. (eds.). *Child Abuse, Domestic Violence, and Animal Abuse*, Purdue University Press, West Lafayette, Indiana, pp. 38-49.

Serpell, J.A. (2000) Domestication and history of the cat. In Turner, D.C. and Bateson, P. (eds.). *The Domestic Cat: the biology of its behaviour*, 2nd edn., Cambridge University Press, Cambridge, pp. 180-192.

Serpell, J.A. (2002) Anthropomophism and anthropomorphic selection – beyond the "cute response". *Society & Animals* **10**, 437-454.

Sharpe, M.S. and Wittum, T.E. (1999) Veterinarian involvement in the prevention and intervention of human violence and animal abuse: a survey of small animal practitioners. *Anthrozoös* **12**, 97-104.

Siegford, J.M., Walshaw, S.O., Brunner, P., and Zanella, A.J. (2003) Validation of a temperament test for domestic cats. *Anthrozoös* **16**, 332-351.

Sims, V.K. and Chin, M.G. (2002) Responsiveness and perceived intelligence as predictors of speech addressed to cats. *Anthrozoös* **15**, 166-177.

Soennichsen, S. and Chamove, A.S. (2002) Responses of cats to petting by humans. *Anthrozoös* **15**, 258-265.

Spencer, L. (1992) Pet prove therapeutic for people with AIDS. *J. American Veterinary Medical Association* **201**, 1665-1668.

Stammbach, K.B. and Turner, D.C. (1999) Understanding the human-cat relationship: human social support or attachment. *Anthrozoös* **12**, 162-168.

Stewart, M.F. (1999) *Companion Animal Death: A Practical and Comprehensive Guide for Veterinary Practice*, Butterworth Heinemann, Oxford, UK.

Swabe, J. (2000) Veterinary dilemmas: ambiguity and ambivalence in human-animal interaction. In Podberscek, A.L., Paul, E.S. and Serpell, J.A. (eds.), *Companion Animals and Us: Exploring the Relationships Between People & Pets,* Cambridge University Press, Cambridge, pp. 292-312.

Triebenbacher, S.L. (2000). The companion animal within the family system: the manner in which animals enhance life within the home. In Fine, A. (ed.), *Handbook On Animal-*

Assisted Therapy: Theoretical Foundations and Guidelines for Practice, Academic Press, San Diego, USA, pp. 357-374.

Turner. D.C. (2000a) Human-cat interactions: relationships with, and breed differences between, non-pedigree, Persian and Siamese cats. In Podberscek, A.L., Paul, E.S. and Serpell, J.A. (eds.), *Companion Animals and Us: Exploring the Relationships Between People & Pets,* Cambridge University Press, Cambridge, pp. 257-271.

Turner, D.C. (2000b) The human-cat relationship. In Turner, D.C. and Bateson, P. (eds.). *The Domestic Cat: the biology of its behaviour,* 2nd edn., Cambridge University Press, Cambridge, pp. 194-206.

Turner, D.C. and Rieger, G. (2001) Singly living people and their cats: a study of human mood and subsequent behaviour. *Anthrozoös* **14**, 38-46.

Turner, D.C., Rieger, G, and Gygax, L. (2003) Spouses and cats and their effects on human mood. *Anthrozoös* **16**, 213-228.

Turner, D.C., Feaver, J., Mendl, M. and Bateson, P. (1986) Variations in domestic cat behaviour towards humans: a paternal effect. *Animal Behaviour* **34**, 1890-1892.

Turner, D.C. and Stammbach-Geering, K. (1990) Owner assessment and the ethology of human-cat relationships. In *Pets, benefits and practice,* Waltham Symposium 20, BVA Publications, London, pp. 25-30.

Vigne, J-D., Guilaine, J., Debue, K., Haye, L. and Gerard, P. (2004) Early taming of the cat in Cyprus. *Science* **304**, 259.

Warner, R.D. (1984), Occurrence and impact of zoonoses in pet dogs and cats at US air force bases. *American J. Public Health* **74**, 1239-1242.

Zasloff, R.L. (1996) Measuring attachment to companion animals: a dog is not a cat is not a bird. *Applied Animal Behaviour Science* **47**, 43-48.

Zasloff, R.L. and Kidd, A.H. (1994a) Attachment to feline companions. *Psychological Reports* **74**, 747-752.

Zasloff, R.L. and Kidd, A.H. (1994b) Loneliness and pet ownership among single women. *Psychological Reports* **75**, 747-752.

Chapter 4

BEHAVIOUR PROBLEMS AND WELFARE

Sarah E. Heath
Behavioural Referrals Veterinary Practice, 11 Cotebrook Drive, Upton, Chester CH2 1RA, UK

Abstract: Behavioural changes are an important indicator of feline welfare. An understanding of normal behaviour, and the constraints that are placed upon it by the domestic environment, is essential if the welfare of the cat is to be safeguarded. The presentation of feline behavioural problems to veterinary practices and behaviour counsellors is increasing, and highlights the need for owner education in feline behaviour and for veterinary understanding of the close links between behaviour and disease. Consideration is given to the incidence and nature of behaviour problems in the feline population, the influence of natural behaviour on reported problems, and the interplay between behaviour and medical conditions. The welfare implications of some of the most commonly presented feline behaviour problems are discussed.

1. INTRODUCTION

1.1 The Changing Attitude to Feline Behaviour Problems

The history of the relationship between humans and cat is a complex one, affected by many diverse influences such as religion and witchcraft. Cats have been worshipped as goddesses on the one hand and reviled as agents of the devil on the other (Serpell 2000). The very same feline qualities that appeal to some people make them abhorrent to others.

In recent times cats have enjoyed an increase in popularity in many, though not all, countries (see Chapter 3), and many veterinary practices in the United Kingdom report that feline patients are numerically more prevalent than canine ones. In a world where people are in need of companionship, cats offer them an outlet for nurturing while retaining a degree of independence and not requiring too much commitment.

The modern cat is primarily kept as a companion (see Chapter 3) and its traditional role as a rodent controller has significantly reduced, if not disappeared, in many quarters. Indeed, many of the aspects of natural feline behaviour, that made the cat so useful to society in the past, are now redundant within the context of its relationship with humans. As a result, misunderstandings and misinterpretations abound. For example, owners often find it difficult to reconcile their pet's companionship role with its instinctive hunting behaviour, so aspects of feline activity that once resulted in them being valued are now reported as problematic. Elements of feline social behaviour can also be interpreted as being undesirable: cats were once expected to be unsociable and aloof, but such characteristics are now reported as behaviour problems.

There has also been an increase in the level of reporting of behaviour problems, with more owners seeking assistance in altering their pet's behaviour. Veterinary practices are increasingly being consulted about feline behaviour, and referral to behaviourists is no longer considered unusual. While there is still a predominance of canine work within the field of companion animal behaviour counselling in the United Kingdom, there are many behaviour practices where work with cats makes up a significant percentage of their workload.

1.2 The Effects of Behaviour Problems on Relinquishment to Shelters and Euthanasia

While reporting of behaviour problems has certainly increased in a feline context, there are many cases where professional help is not sought, and the options of relinquishment to a shelter or euthanasia are seen as the most appropriate course of action (see Chapter 5). The Cats Protection is the largest feline welfare charity in the United Kingdom, and re-homes approximately 60,000 cats and kittens every year. In its annual report (Cats Protection 2003), the charity listed the reasons for 6,089 cats being brought to 11 of their centers (one main reason was given per cat), and behavioural problems accounted for 7% of cases (Table 1).

Of the 456 cats relinquished for behavioural reasons, 36% exhibited problems of cat to cat aggression, 17% house soiling, 15% aggression to people, 12% showed general nervous or fearful behaviour, 10% problems of integration with a dog in the household, 1% scratching of furniture and 9% had a variety of other behavioural problems.

Table 1. Reasons given by owners for the relinquishment of their cats (n=6,089) to Cats Protection shelters.

Reason for relinquishment	Percent of cases
Stray/abandoned	32
Owner circumstances	19
Kittens	14
Cat transfer within Cats Protection	9
Problem behaviour	7.5
Owner can't cope	5
Allergy/asthma	5
No reason	3
Too many cats	2
Owner pregnancy/child	2
Cat ill/pregnant	1
Other	0.1

In another survey, by the British charity The Blue Cross (Blue Cross 2003), of 3,021 cats from 11 of its rescue centres, 11% were relinquished because of problem behaviour (one reason was given per cat; Table 2). Reasons for relinquishment of cats to American shelters are presented in Chapter 5.

Table 2. Reasons given by owners for the relinquishment of their cats (n=3,021) to Blue Cross shelters.

Reason for relinquishment	Percent of cases
Stray/abandoned	29
Change in owner circumstances	14
Owner moving	13
Problem behaviour	11
Can't cope	8
Allergy/asthma	7
Unwanted litter	4
Financial problems	2
Not allowed to keep	1
Feral	1
Other	10

The relationship between feline behaviour problems and euthanasia is less well researched, although a survey conducted in 1998 into the reasons for euthanasia of animals in veterinary practices in England found that 1% of 385 cats were euthanased because of behavioural problems (Edney 1998).

2. THE MOST COMMONLY REPORTED BEHAVIOUR PROBLEMS

The Association of Pet Behaviour Counsellors (APBC) is an organization, based in the United Kingdom, that represents an international network of experienced and appropriately qualified pet behaviour counsellors. They treat behaviour problems in dogs, cats, birds, rabbits, horses and other pets, who are referred to them by veterinary surgeons. In the United Kingdom, the most regularly published information about behaviour problems in cats is produced by the APBC in its annual review (Association of Pet Behaviour Counsellors 2003). The data in the review are derived from clinical cases that have been referred to a selection of members of the association; therefore they represent a limited and somewhat biased population. However, they give an overall impression of the incidence of feline behaviour problems, the types of problems and the breeds in which these problems are reported. Of 66 feline cases (32 male and 34 female, 97% neutered), the most commonly reported reason for referral was indoor marking behaviour (25%). This category included the deposition of urine, either through urine spraying or squat marking, and the strategic deposition of faeces, which is termed middening. Aggression towards people accounted for 23%, while aggression to other cats accounted for 13% of the cases reviewed. Other problem behaviours, such as difficulties with house training, attention seeking, and self-mutilation made up 12%, 11% and 6% of the referred population respectively. The average number of problems per cat was 1.7 (there was no difference between males and females)

The majority of cats referred to the APBC for behavioural problems were domestic shorthairs (57%). This is likely to be a reflection of the predominance of domestic shorthairs in the feline population of the United Kingdom. The two most commonly referred oriental breeds were the Siamese (11%) and Burmese (11%), which are also quite well represented in the general cat population.

The discrepancy between the incidence of feline behaviour problems in the referred population and in the general cat population was highlighted by research at Southampton University (Bradshaw *et al.* 2000; Casey 2001). Clinical data collected from 83 cats treated by a veterinary behaviourist at the University's referral clinic were compared with data obtained from a general population survey carried out in the United Kingdom, and data from a published national diagnostic review from the United States. The general population survey was conducted by door to door enquiries in two locations, one a suburban area in Hampshire and the second a rural area in Devon. In total, 90 owners completed questionnaires detailing the occurrence of unwanted behaviour in 161 cats. The national diagnostic review was based on the analysis of postings on an "ask the expert" website produced by the Discovery Channel.

The results suggested that certain problems, particularly those of feline house-soiling (both elimination and marking), and aggression (directed toward people or other household cats), were over-represented in the referral clinic population. In contrast, fearful and avoidance-related behaviour, whether toward unfamiliar people or other cats, was not as frequent as would be predicted from the incidence in the general population, and scratching behaviour was also under-represented in the referral population.

Explanations for the discrepancies between the data from the referred population and from the two general cat populations are probably related to the potential impact of these different behaviours on the owners, and also to the level of knowledge amongst the general cat-owning population about behavioural therapy for cats. Problems of house soiling and inter-cat aggression can be difficult to live with, and a high proportion of owners will be driven to seek help in order to resolve these issues. In contrast, cats that run and hide from fear-inducing situations do not pose any direct challenge to their owners and are often dismissed as being "a bit nervous", so their owners will be less likely to seek help.

3. NORMAL FELINE BEHAVIOUR

3.1 Introduction

As mentioned previously, the change in the role of the cat from that of an independent hunter and rodent controller to that of a companion animal, can lead owners to be less tolerant of their cat's natural behaviours and perceive them as problematic and undesirable. In some cases, failure to allow the expression of normal feline behaviour and communication leads to behavioural change in the cat, that is problematic both to the cat and its owner. In other cases, the cat is not troubled by its behaviour but the behaviour is inconvenient to its owner. In the former situation, resolution of the behavioural problem is clearly important for the welfare of both cat and owner, and a full investigation is necessary to identify the causes of this behaviour and to formulate a treatment plan. This may well involve modifying the cat's social and physical environment, with the aim of accommodating its natural feline behaviours and of treating the cat's problematic behaviour to remove detrimental aspects, such as over-grooming, inappropriate elimination or indoor marking, from its repertoire. In the second situation, the possibility that there will be inappropriate reactions to the problem behaviour, such as punishment, may also raise welfare concerns for the cat. The main aim of intervention is to educate owners so that they can accept certain feline behavioural traits, such as scratching, predation and short duration, high-intensity activity, which cannot be entirely removed from the animal's repertoire.

3.2 Normal Feline Behaviours that are Problematic to Humans

3.2.1 Scratching Behaviours

Although often reported by cat owners as being a problematic behaviour, scratching rarely results in referral to a behaviourist. In most cases, damage to furniture and curtains is either tolerated or minimised by limiting the cat's access to certain parts of the house. The use of scratching posts within the home is usually successful in redirecting the behaviour to specific locations, but an understanding of the ethological basis of scratching is important if owners are to fully understand their cats and cater for their behavioural needs. Scratching is a complex behaviour pattern, which has both functional and communication-related components (Bradshaw 1992). Cats will scratch in order to remove the blunted outer claw sheaths from their front claws, and to exercise the apparatus used to efficiently protract the claws during the hunting sequence. For this reason, all cats have a basic need to scratch and it is essential that facilities are provided to enable them to do so. These may be found in the outdoor environment in the form of tree trunks, fence panels and garden sheds, but when domestic cats have restricted outdoor access, it is necessary for the facilities to be made available within the home (see Chapter 7). Some owners go to great lengths to provide scratching areas, for example by covering the bottom half of doors with carpet or hessian material, but the majority of owners rely on commercial scratching posts to fulfil this function. One major drawback of these products is their restricted height: functional scratching requires enough height for the cat to scratch at full stretch and thereby gain sufficient purchase on the surface to scratch effectively. If suitable facilities are not provided, cats will scratch on any convenient and accessible surface within the home; backs of furniture and draping curtains are often targeted.

Scratching is also used as a form of communication, both through the deposition of scent signals from specialised glands on the paws and also through the creation of vertical striations in the scratching surface (Bradshaw 1992). The exact meaning of these signals has not been fully elucidated, but when a domestic cat begins to scratch in behaviourally significant locations within the home, such as near points of entry and exit or on specific walkways, it is important to investigate this behaviour in some detail. The use of scratching as a marking behaviour is normal along regularly used routes in the wider territory (Feldman 1994) but when it occurs within the home, which should be regarded as a safe and secure core territory, it crosses the boundary between normal and problematic behaviour and raises concerns about whether the cat is feeling insecure. When scratching and urine marking within the home are exhibited concurrently in multi-cat households, investigation of social interactions, including passive and active aggression between cats, should be carried out (Overall 1997).

3.2.2 Predatory Related Activity

Cats are the most specialised living carnivores (Kitchener 1991) and, while they are not always very efficient in the capture of their prey (Turner & Meister 1988), they are finely tuned to respond to sensory stimuli which signal its presence. Since cats use auditory and visual signals to locate suitable prey items, their sharpened responses to high-pitched sound and rapid movement can sometimes be perceived as problematic by owners. The cat chasing the pen top as its owner attempts to write a letter may be amusing, but when cats hunt wildlife and bring it home to the safety of their feeding lair, many owners find this behaviour distasteful and distressing. There is nothing abnormal about a cat killing small rodents, birds and even insects, but owners expect cats to prefer the proprietary food that they provide. When hunting continues despite an adequate level of feeding, owners are dismayed. After all, there is an implication of pleasure when hunting continues even though it is not necessary for survival.

Owners need to understand that the motivation to hunt and the sensation of hunger are distinct (Bradshaw 1992) if they are to avoid forming negative associations with their cats. Even a well-fed cat will still be motivated to hunt if the correct sensory stimulation is present; this is understandable in terms of feline survival. Cats are solitary hunters (Fitzgerald & Turner 2000), and if hunger were the trigger for the predatory response, it would mean that cats would already be slightly debilitated at the onset of the hunting sequence. While some studies have suggested that the high percentage of domestic cats with identifiable prey items in their stomachs reflects a high success rate in predatory terms (Spittler 1978), it must be remembered that this is a weak measure of success since it does not take into account the number of unsuccessful attempts that were made in the process. Indeed, research by Corbett (1979) into the hunting success of pet and feral cats showed that the cats had a success rate when hunting rabbits as low as 17%. At such rates, hunger-induced hunting would put cats in a potentially vulnerable position and run the risk that successful acquisition of prey will be jeopardised by physical weakness. Instead, cats are motivated to hunt by the presence of specific sensory cues, and they will respond to these cues regardless of their level of hunger. Hunger can serve to intensify the effort put into the hunting sequence, but even if a cat has just eaten a full meal it will not ignore the signals of high-pitched sound, rustling and rapid movement. In consequence, cats will capture and even kill potential prey that they have no intention of consuming. While such surplus prey would be eaten by other animals in the wild, there is an inevitable sense of waste in the domestic context. This can have a negative impact on the way cats are regarded in general, and cause some owners to have mixed feelings toward their cats.

The effects of feline predation on the wildlife population have been widely debated (see Chapters 3, 5 and 6), but education of cat owners is possible in order to reduce them; this is important for the welfare of both cats and wildlife. Keeping cats indoors at times when prey are most likely to be active will be beneficial, so dawn and dusk curfews are advisable (see section 3.2.3). However, simply limiting access to the prey population is not enough, as many cats seem to have a need to engage in predatory behaviour. Provision of opportunities for the expression of predatory behaviour through play, both for kittens and adults, is therefore an essential aspect of cat ownership (Figure 1). Research has shown that objects which are mobile, have complex surface textures and mimic prey characteristics are the most successful at promoting play (Hall & Bradshaw 1998). There are significant welfare implications when opportunities for expressing predatory behaviour through play are not provided, as the risk of problems as diverse as aggressive behaviour toward people and cats in the household, and obesity (see section 5.3), may increase.

3.2.3 Feline Activity

Another aspect of normal feline behaviour that can cause problems in the domestic environment is the style and timing of feline activity. Concern over the effects of feline predation on the wildlife population has led to the recommendation from organizations in the United Kingdom, such as Cats Protection, that cats should be kept indoors from before dusk until after dawn (reducing the risk of road traffic accidents is another reason for this advice, see Chapter 7). This recommendation is based in part upon the belief that cats are naturally crepuscular (with peaks of activity at dawn and dusk), although the research is inconclusive (Fitzgerald & Turner 2000) and some studies have suggested that the domestic cat has developed a more diurnal pattern of activity. Nevertheless, high levels of activity around the times of dawn and dusk are often reported by owners and may be considered to be incompatible with domestic life.

Figure 1. Opportunities for the expression of predatory behaviour through play should be provided for both kittens and adult cats.

This problem can be partially alleviated by the effects of conditioning, whereby the cats learn to be more active at times when owners are available to interact with them. But whatever time of day the activity occurs, the natural feline pattern of short bursts of high energy-consuming behaviour interspersed with significant periods of rest can lead to problems. Owners will often report that their cat has periods of 'madness' during which it dashes around the house at high speed and appears to be having some kind of fit. Cats normally eat small but frequent meals (see Chapter 9), which fuel frequent but short bursts of intense activity interspersed with prolonged periods of sleep. When cats have access to the outdoors, such a pattern can be quite effectively maintained with bursts of activity, such as climbing trees or hunting for prey, followed by extensive rest periods. However, when the cat is kept indoors or has restricted access to the outside, some of the energy-consuming sessions will inevitably take place inside the house. In multi-cat households, the synchronisation of activity by several residents can intensify the perception of madness, and can also lead to misinterpretation of the cats' intense chase and flee responses as a form of aggression. Providing cats with outlets for high energy-consuming activity at various times during the day will encourage a more even distribution of activity, and make the behaviour more compatible with domestic life. Adjusting feeding regimes so that cats have *ad libitum* access to food, to fuel the intense periods of activity, is also advisable.

4. BEHAVIOUR PROBLEMS

4.1 Introduction

When the constraints of domestic life adversely affect their cat's behaviour, owners may label their cat as problematic and seek professional advice. As some behaviours are detrimental not only to the owner but also to the welfare of the cat, effective investigation and treatment are crucial. The successful treatment of behavioural problems relies largely on an accurate identification of the underlying motivation. During the history taking process, it is important to assess the animal's emotional state, both during the expression of the problem behaviour and in a variety of other contexts and situations. Identification of factors that have predisposed the individual to developing behavioural problems, as well as events that may have initiated the particular behavioural changes, will be important. An understanding of learning theory is necessary in order to identify both intentional and unintentional reinforcers, and once the behavioural motivation has been accurately identified, treatment will involve the removal, where possible, of factors that are initiating and maintaining the behaviour.

4.2 Early Development

Over recent years, considerable attention has been given to the subject of early socialization and habituation in dogs, and a lack of adequate and appropriate exposure to novelty and challenge in early life has been identified as one of the important predisposing factors for a range of behavioural problems (Appleby *et al.* 2001). Fears and phobias in dogs have been shown to be a potential cause of unacceptable behaviour toward people and other dogs, and as a result there has been considerable interest in how such problems can be prevented as well as treated. While feline fears and phobias may not have as direct an effect on humans as canine ones, the influence of early development on feline behaviour is still highly relevant to the welfare of the domestic cat, and the importance of socialization in preventing behavioural disorders is just as important (Hunthausen & Seksel 2002).

Information about the potential for the prevention of unsociable behaviour in cats is largely derived from research by Karsh and Turner (Karsh 1983; Karsh & Turner 1988). Their work showed that feline responses to people, in terms of their reaction to handling and social interaction, could be significantly influenced by the level of human exposure that kittens received in their first few weeks of life. Increases in the time that cats would voluntarily spend in human company, and corresponding decreases in their hesitation to approach unfamiliar people, demonstrated how the amount and frequency of handling could affect feline social behaviour. Related research highlighted the importance of other factors, such as the number of different handlers (Collard 1967), in producing kittens that had the potential to make good pets (see Chapter 3).

Failure to take such findings into account when rearing kittens increases the risk of producing adult cats that are overtly fearful and even aggressive, and while many of these cats are never brought to the attention of companion animal behaviourists, the welfare implications cannot be overlooked. Some cats will establish ways of controlling their fear responses and hide from potential stressors (Figure 2), but for many cats living in a domestic environment, the constraints of that existence only serve to accentuate the shortcomings of the rearing process.

Hiding is an important feline coping strategy (see Chapter 2), and in a study of the behavioural and physiological correlates of stress in laboratory cats, hiding behaviour was negatively correlated with urinary cortisol concentration (Carlstead *et al.* 1993). However, many domestic cats live in environments that fail to provide any opportunity for them to hide (see Chapter 7), and as a consequence fear-related behaviour may seriously decrease their welfare. A failure to understand feline social behaviour can exacerbate the situation, as owners attempt to initiate social interaction with their poorly socialized cats, thereby inducing and perpetuating fear-related behaviours, including aggression. An appreciation of the fact that cats will spend more time interacting with their owner when they, rather than the owner, initiate the interaction will help to improve the cat–owner relationship and consequently the welfare of both partners in that relationship (Turner 2000) (see Chapter 3).

Figure 2. Some cats will attempt to control their fear response by hiding, such as this Siamese cat trying to hide under furniture.

Another aspect of early development which has welfare implications, in terms of its potential influence on the development of problem behaviour, is weaning. During the weaning process, kittens becoming increasingly responsible for initiating the nursing process (Schneirla *et al.* 1966) and there is an increasing element of behavioural independence from the mother. The development of identification of prey and practice of stalking and capturing behaviours prepare the kittens for a predatory existence, and in the wild the behaviour of the mother in bringing back dead and dying prey to the nest is crucial to this process (Kitchener 1991). In the domestic situation the aim of the weaning process is somewhat modified, being a transition from milk to proprietary cat food; however, weaning continues to be important in terms of fostering independent adult feline behaviour patterns. Neville and Bessant (2000) have suggested that the weaning process teaches kittens to deal with frustration. They report that inappropriate weaning, which fails to mimic the queen's ability to frustrate her offspring by giving signals of non-reward for previously rewarded behaviour when they approach for milk, may actually predispose kittens to problems of aggression in adulthood. Furthermore, they suggest that as a result of the extra handling that the process involves, hand-reared kittens may retain attractive, affectionate and dependent infantile behaviours when adult, but they may also alternate rapidly in mood and become highly reactive and aggressive towards their owners.

4.3 House Soiling

One of the most commonly referred behavioural problems in cats is house soiling (Association of Pet Behaviour Counsellors 2003), yet the incidence of these behaviours in the feline population does not appear to reflect this (Bradshaw *et al.* 2000). One of the reasons for this discrepancy may be that the potential impact of such problems on human quality of life is significant, leading to an increased likelihood of referral, but the welfare implications for the cat also need to be considered.

House soiling problems can be divided into those which involve inappropriate elimination, and those which involve the use of urine or faeces as markers within the home environment. When considering cases in either category, it is important to assess the underlying motivation and to understand the influence of natural behaviour. Obviously, examination of any potential medical factors is a priority since physical disease can not only cause, but also exacerbate, house soiling problems (Horwitz 2002). Differentiation between elimination and marking will require accurate history taking, and information about the cat's posture together with details about the location, frequency and size of the deposits will be useful in determining the motivation.

Elimination problems are often associated with a decreased use of litter facilities or reluctance to venture outside, and the selection of secluded and private locations for the deposits. Inappropriate marking behaviour is most commonly associated with small, frequent deposits of urine onto vertical surfaces, with cats adopting a characteristic "spraying posture". However, it is also possible for cats to mark with urine and faeces on horizontal surfaces so information about their behavioural responses in other contexts will also be useful.

4.3.1 Inappropriate Elimination

Cats' fastidious attention to hygiene is one of the reasons cited for their increasing popularity as a pet. In the wild, cats will bury urine and faeces in order to reduce odour and limit the risk of disease and parasite spread (Beaver 2003). For cats that spend the majority or all of their time indoors, this natural behaviour is readily transferred to litter facilities, while cats with access to the outdoors will eliminate in soil or other substrates. When cats begin to deposit urine and faeces in unacceptable locations, it is important to consider why the intended latrine facilities are not being used and why the new location is being targeted. Treatment involves identifying the underlying motivation and then attempting to remove or control it, while eliminating any habitual component to the behaviour and re-establishing appropriate latrine associations.

In the case of cats that have previously eliminated away from the home, any factors that may make the cat reluctant to venture outside, such as social tension between cats in the neighbourhood, should be considered. Other fear-related factors, which make the outside world threatening, should also be examined. The act of elimination can make the individual vulnerable, so if there is insufficient vegetation for the cat to hide in during the act, it may seek a dark and secluded location within the house as an alternative. Such behaviour suggests that the cat feels safe within the home environment, so if the owner uses punishment in an attempt to deal with the problem, the resulting decrease in the cat's feeling of security can have important welfare implications.

In cases where cats are confined to an indoor environment, relationships between cats in the household should be reviewed in order to identify any possible social influence on the use of the facilities (see section 4.4). The general rule concerning the number of litter trays in multi-cat households is to provide one litter tray per cat and one spare (Hetts & Estep 1994). However, it is not uncommon for owners to expect cats to share latrines at a far lower ratio, and this has implications both in terms of the cleanliness of the facilities and in terms of social compatibility. The physical accumulation of waste within the tray, and the resulting odour that it produces, are significant aversive stimuli which can lead to cats refusing to use the facilities and seeking other locations. In some cases the tray may be visibly dirty, and deposits of faeces and urine may even make it difficult for the cat to enter the tray, but in others the tray may appear to be clean on superficial inspection but a slight disturbance of the litter substrate may release a strong odour of ammonia. In multi-cat households covered litter trays are often selected in order to provide additional privacy, but the build-up of unpleasant odours within the covered tray may be considerable (Beaver 2003). In view of the cat's naturally fastidious nature, the need to eliminate in the presence of excreta, especially from other cats, is likely to be stressful, so owners should ensure that their cats have easy access to clean facilities at all times (see Chapter 7). Scooping out urine and faeces once or twice a day should be considered a minimum, and emptying the tray completely and replacing the litter with a fresh supply is recommended weekly (Landsberg et al. 2003b). The use of boiling hot water, rather than disinfectants which can interfere with the odour identity of the latrine, is advisable.

In addition to considering the number of facilities that are available to cats within the home, it is also important to consider the appropriateness of those facilities in terms of substrate. Changes in litter type can be associated with the onset of inappropriate elimination (Crowell-Davis 2001a), and one author suggests that 50% of cats will stop using the litter tray if there is an abrupt change of substrate (Beaver 2003). A number of sensory components of litter substrates, such as texture and odour, may influence their use (Halip 1994), so in-depth history taking is important in every case. Many cats will refuse to use a tray due to the presence of an aversive substrate, and since there is a degree of individual preference involved in the selection of substrates, a limited choice can pose problems. Preference tests suggest that the most widely-accepted substrate is a soft, easily raked, clumping litter (Borchelt 1991). Selection of a substrate which is easily raked makes sense in terms of the cat's natural desire to dig and bury during the elimination process. However, cats can also develop individual aversions to particular substrates as a result of negative associations, such as pain from medical conditions, so while the substrate may appear to be quite suitable, it may not be acceptable to the individual cat. Offering a choice of litters in a number of trays can sometimes help in the diagnostic process.

Unacceptable location of litter facilities is another factor that can contribute to feline elimination problems, and offering choice is once again a desirable approach. Various factors may influence the cat's perception of a latrine location, and history taking will be important in identifying possible reasons for the rejection of the location. The placing of trays under staircases and beside household appliances should be reconsidered, as cats are very sensitive to auditory stimuli. In households where dogs or small children are present, the location of the tray needs to be sufficiently inaccessible and private so that problems of intimidation are avoided. Tension between cats in the same household can also reduce accessibility (Overall 1997), and in these cases treatment involves the provision of sufficient numbers of latrines in a choice of locations.

4.3.2 Indoor Marking

Although urine spraying is the most common form of indoor marking to be referred for behavioural therapy, cats can also mark their territory by squat marking with urine (Borchelt 1991), by depositing faeces in prominent places (middening) (Simpson 1998), and by scratching to leave both a visual and an olfactory message.

In the wild, cats use the central part of their territory for the activities of eating, sleeping and playing, and marking behaviours are only performed at the periphery of the territory or in the home range beyond (Bradshaw 1992).

It is therefore important, in cases of inappropriate marking, to understand why the cat is marking within its core territory and to try and assess its emotional state, both when marking occurs and in other contexts. In the past, marking was associated with confident individuals trying to assert control, but it is now thought that marking is the act of an insecure cat. The presence of scent signals, that identify the location as their own, appears to result in increased confidence and, whilst spray marking is important for communication with other cats when it is used outdoors, it is thought that indoor spraying is an important method of self-reassurance (Hart & Hart 1985).

In contrast to elimination, marking behaviours can occur daily at a very high frequency, and their potential to damage the cat-owner relationship can be correspondingly increased. Misunderstanding the purposes of marking in feline communication can lead to the use of punitive techniques in an attempt to reform the behaviour, and in turn this can decrease the cat's feeling of security and increase its motivation to mark. The influence of multi-cat households on the rate of indoor marking is not clear, and while it has been suggested that the incidence of spraying is proportional to the number of cats in the household (Beaver 2003), it has also been stated that the severity of the problem is likely to be worse in single-cat households (Hunthausen 2000).

Urine marking is a normal form of feline communication and, while it can be associated with sexual behaviour, it is usually the reactional or emotional form of the behaviour that is presented as a problem in a veterinary context. In such cases, territorial behaviour and anxiety are the most common underlying motivations (Beaver 2003), and cats deposit urine in response to stressors, be they real or perceived. According to research from the United States the most commonly targeted items include furniture, walls or windows at the periphery of the home (where cats have visual access to the outside), electrical appliances and novel items (Pryor *et al.* 2001). Analysis of the spatial distribution of the behaviour can be very important in establishing the underlying motivation, so asking owners to draw plans of their property and note the marking incidents in chronological order can be extremely beneficial. Some of the common causes of indoor marking include agonistic interactions with cats outside the home, aggressive interactions with cats in the home, limited access to the outdoor environment, moving home, introduction of novel objects into the household and interactions with the owner (Pryor *et al.* 2001). The welfare implications of exposure to these stressors should be taken into account, especially where the cat has a limited ability to either predict that exposure or control it when it occurs. In some cases, re-homing of the cat to another environment may be indicated.

One of the aims of treatment is to break the habitual component of the behaviour, which results from the motivation to top up scent signals and keep them fresh (Bradshaw 1992). This involves effectively cleaning the deposits in order to remove the scent (Heath 2004). Inappropriate use of cleaning agents which contain ammonia and chlorine can lead to confusion, since the scent is suggestive of the presence of another cat and the resident cat may be induced to mark again. In many cases, owners interpret this behaviour as being spiteful, since the cat has targeted a freshly clean location to mark. At this point, the breakdown in communication between cat and owner can become severe and the likelihood of punishment being employed is increased.

Where specific stressors can be identified, treatment aims to remove these from the cat's environment or at least limit their potential influence. Specific behavioural therapy to assist in the successful integration of housemates, and desensitisation and counter-conditioning techniques in relation to identifiable fear-inducing stimuli, can all be beneficial. However, many of the cats presented with problems of indoor marking have been performing the behaviour for some considerable time prior to referral, so the use of medication and/or pheromone therapy in conjunction with behavioural modification may be necessary (Horwitz 2002; Beaver 2003). The aim of such adjunctive treatment should always be to support the behavioural programme in the short term and, while medication and pheromone therapy can be extremely useful in reducing anxiety, it is also important to rectify the environmental challenges that led the cat to mark indoors.

4.4 Multi-cat Households

Cats naturally live in groups of related individuals (Bradshaw 1992) and feline social relationships are often enduring in nature (Beaver 2003). The longer the individuals are together the stronger the bond becomes, and this is especially true when the relationship starts early in life. Siblings kept together are observed to spend more time engaging in social interaction than unrelated cats (Bradshaw & Hall 1999) and their relationship is often characterised by behaviours such as mutual grooming, hunting close to each other and sleeping together. Unrelated cats tend to eat one at a time if there is only one feeding station available, or select to eat at well-separated bowls where these are on offer; in contrast, siblings often share a food bowl (Bradshaw & Hall 1999). Keeping sibling groups together in households would appear to be in line with natural feline behaviours, and some re-homing centres strongly advise the homing of kittens in sibling pairs. In feral colonies, cats are usually resistant to the introduction of newcomers. Acceptance of a new cat into a feral colony is a slow process, and the outsider is initially very much on the periphery of the social group (Crowell-Davis 2001b; see Chapter 1).

In contrast, the practice of keeping a number of unrelated individuals is commonplace within the domestic environment (Figure 3), and owners often expect resident cats to accept newcomers. Of course, one of the major differences between wild and feral colonies and the domestic environment is the rate of neutering, and this has been shown to influence social behaviour toward cats from outside the social group.

Figure 3. Siblings are observed to spend more time engaging in social interaction than unrelated cats, but keeping a number of unrelated cats in households is commonplace. (Courtesy of Penny Bernstein).

When studies of the ethology and population dynamics of neutered colonies are compared with similar studies in breeding colonies, the most obvious differences relate to the absence of maternal and sexual behaviours, but another behaviour which is reduced in the neutered colonies is aggression to members of neighbouring groups (Bradshaw 1992). It is perhaps this alteration in social behaviour which explains the relative success of many multi-cat households, but keeping cats in unrelated groups is not without its problems. When owners attempt to increase the size of their feline household through the introduction of a newcomer, hostility is common. Limiting the problem through careful attention to the introduction process is recommended, and when aggression has already occurred it is advisable to separate the cats immediately and embark on a gradual, controlled introduction process following a period of separation (Heath 2002). The aim is to give time for the resident cats to come to terms with the arrival of a new scent profile, and the newcomer to establish itself in the new territory before the face-to-face introductions begin (Beaver 2003).

Whilst feline hostility is almost to be expected in reaction to the introduction of a newcomer, breakdowns in the stability of feline social groupings that have worked well for considerable periods of time, can be very distressing for owners and cats alike. These social disruptions are often triggered by a challenge to the scent profile of the social group, for example through an indoor cat gaining unexpected access to the outdoors or a resident leaving the group to attend a veterinary practice for medical attention (Heath 2002). Temporary separation, during which the scents of the two cats can be reintroduced in the absence of any visual contact, is recommended (Crowell-Davis et al. 1997) and should be followed by a gradual reintroduction along the same lines as when introducing a total stranger.

Keeping cats in socially incompatible groupings can be very stressful, and if the environment does not offer sufficient opportunities for retreat and hiding the possibility of re-homing may need to be seriously considered. Even where gradual introduction results in a state of reasonable toleration, and the multi-cat household appears to function at a level that is acceptable for all of those concerned, it is still important to remember that there may be multiple social groupings within the one household and the provision of essential resources needs to reflect this. Sufficient water and feeding stations, sleeping locations and latrines are essential if social stress is to be avoided, and environmental management to provide cats with sufficient access to three-dimensional space and to appropriate hiding refuges, is central to the success of a multi-cat household (see Chapter 7). The fact that cats will come together in order to feed is often interpreted as a sign of social closeness, but the fact that food is provided by the owner in one location within the home dictates this temporary acceptance of physical proximity. Limited availability of drinking stations, on the other hand, may lead to a decrease in water intake with resulting medical implications (see section 5.2.1), and lack of access to latrines can result in house soiling issues, as discussed earlier.

Even where social compatibility is not an issue, the provision of sufficient resources to allow each cat access when necessary is still important. Research suggests that dominance hierarchies are not a feature of feline society (Bradshaw & Lovett 2003); one of the implications of this is that access to resources is not controlled by hierarchical rules. The concept of waiting for turns or deferring to higher-ranking individuals is probably irrelevant (see Chapter 1 for another view), so the provision of an adequate supply of essential resources is a major factor in the prevention of social stress within a multi-cat household.

5. HEALTH, DISEASE AND BEHAVIOUR

5.1 Introduction

A very important dimension of behavioural work is the interaction between physical health and behavioural symptoms. Changes in behaviour are often the first signs which alert an owner that their pet is unwell. An understanding of the relationship between behaviour and disease is essential for anyone working in the field of feline behavioural problems. A failure to make the connection between disease and behaviour has significant welfare implications, not only because of the potential for the physical disease to be left untreated, but also because the behavioural symptoms may worsen. A thorough medical investigation of the presenting behavioural problems is therefore essential, so it is important that veterinary support is sought where non-veterinarians are involved. The use of a referral system, where cats are examined by a veterinary surgeon prior to consultation with a behaviour counsellor, is recommended.

5.2 Behavioural Manifestations due to Underlying Medical Causes

Two of the most common behavioural problems that are linked to underlying medical causes are house soiling and over-grooming but other symptoms, such as aggression to people or other cats, can also occur as a consequence of physical disease, for example conditions causing pain. Behavioural changes, such as irritability and hyperactivity, associated with hyperthyroidism in elderly cats are often noticed in the early stages of the disease when other medical symptoms are less obvious. When owners report rapid and significant alterations in their pet's behaviour, neurological conditions, endocrine disorders, diseases affecting the blood flow to the central nervous system and metabolic conditions affecting the levels of circulating toxins should all be considered.

5.2.1 Feline Lower Urinary Tract Disease

The interplay between behaviour and feline lower urinary tract disease is extensive, and behavioural factors can play a significant part in the aetiology of the condition as well as its presentation and treatment. The importance of the relationship between feline welfare and disease is illustrated by the aggravation of symptoms of interstitial cystitis by stress, mediated through the effect of increased sympathetic activity on bladder function (Buffington *et al.* 2002). A recent questionnaire-based study considered whether any environmental or behavioural factors, particularly those that could be considered potentially stressful, were associated with feline idiopathic cystitis (FIC) (Cameron *et al.* 2004). The study compared 31 cats with FIC with 24 cats, living in the same households, that did not have FIC. The two groups were also compared with a control population of 125 clinically healthy cats. Several stress factors were found to be associated with FIC; the most important factor was living with another cat with which there was conflict. The study concluded that stress may be implicated in some cases of FIC.

The most obvious presentation of feline lower urinary tract disease, in behavioural terms, is the deposition of urine in inappropriate places. This can result both from a lack of control over urination, leading to the frequent deposition of small amounts of urine throughout the house, and a learned negative association between the pain of urination and the location of the litter tray, leading to urination in other locations (Horwitz 2002). In addition, FIC can lead to the less obvious presentation of over-grooming, where cats remove fur specifically from the lower abdomen and medial thighs, presumably due to pain from the lower urinary tract.

In addition to stress, other factors such as limited water intake and dry diet consumption have been clearly established as risk factors for the development of FIC (Walker *et al.* 1977; Buffington & Chew 1998; Buffington *et al.* 1999); therefore, behavioural and environmental factors should be investigated as part of the cat's medical history if it is to receive appropriate veterinary treatment. Treatment not only involves the appropriate use of medication, but also the implementation of behavioural and environmental strategies designed to reduce stress and improve access to the all-important resources of drinking water and litter trays (Gaskell 2004). In single-cat households, issues such as access to the outdoors, the presence of aggressive cats in the environment and lack of adequate provision for normal feline behaviours may need to be considered. In multi-cat households, the social compatibility of cats within the home and the potential for competition over important resources, such as water stations and litter trays, should also be taken into account (Cameron *et al.* 2004). If one cat is preventing another from accessing a limited number of water stations, this could significantly decrease water intake, and urinary retention may develop if there is competition over access to litter trays. The household may have to be reorganized to provide safe areas for all feline residents. The use of video recordings of areas frequented by the cats and house plans can be very useful to determine where such areas should be established.

5.2.2 Dermatological Conditions

When cats present with hair loss, it is always important to consider common dermatological conditions, such as flea allergic dermatitis, before assuming that the hair loss has purely a behavioural cause. In some cases, the involvement of a veterinary dermatologist may be advisable. As mentioned previously, feline lower urinary tract disease should be considered when the distribution of hair loss is consistent with lower abdominal pain. When hair loss is combined with behavioural changes and over-sensitivity of the skin, the complex medical condition of feline hyperaesthesia syndrome should feature in the list of differentials (Landsberg et al. 2003a). Diagnosis of this syndrome can sometimes be difficult, but the presence of rippling skin in association with minimal stimulation is one of the characteristic signs. Behavioural symptoms can vary from low-grade irritability to overt aggression, and signs of frustration, including mounting of inanimate objects and dashing uncontrollably around the house, are often evident. Successful treatment of feline hyperaesthesia syndrome often involves medication with tricyclic antidepressants or serotonin reuptake inhibitors, but potential stressors within the cat's environment should also be considered in order to offer a long-term approach to controlling the condition. Cure is seldom achievable and owners need to be aware that symptoms are likely to reoccur when the cat is under stress.

5.2.3 Pain

The development of painful foci can cause the sudden onset of aggressive behavioural patterns in a cat that has been previously good-natured, but pain can also be insidious in its onset and the behavioural consequences can be equally subtle (see Chapter 8). For example, chronic orthopaedic injuries and associated arthritic changes can be responsible for a gradual decrease in the cat's toleration of being handled. Dental pain has also been associated with changes in feline behaviour. Indeed, many owners report that a behavioural change for the better is one of the most noticeable effects of the dental procedure, but that it is not until the procedure has been carried out that they realise how much their cat's behaviour had deteriorated (J. G. Robinson, personal communication).

5.3 Behaviour and Obesity

In many Western countries, human obesity is a condition that is reaching epidemic proportions, and the incidence of this condition in the feline population in these countries is also a cause for concern. Surveys of dogs and cats presented to teaching hospitals or private practitioners indicate that approximately 25 % of cats are overweight. (Hand *et al.* 1989; Markwell *et al.* 1994). In common with the approach used in human medicine, there has been an increasing awareness of the need to incorporate lifestyle changes into a feline obesity management programme (Dehasse *et al.* 2001), as attempts to deal with obesity by dietary management alone may prove unsuccessful. It is often alterations in the cat's behaviour, such as increased inactivity and inability to jump, that first alert owners to their cat's increasing bodyweight, and this restriction of normal behaviour that leads owners to seek assistance. The medical implications of obesity, in terms of acting as a risk factor for a number of physical conditions, make it an important welfare issue (see Chapter 9), but the behavioural implications of excessive bodyweight should also be considered. Obese cats may be unable to engage in predatory activity, to climb to high vantage points and observe their territory, or to groom all areas of their body, and such limitations to their natural behavioural repertoire can have serious implications in terms of their mental health. Improving the balance between energy input and expenditure is the key to successful weight loss, so in a feline context an understanding of normal feeding behaviour (see Chapter 9) and natural activity patterns is essential if this is to be achieved. Educating owners about the need for their cats to receive small frequent meals and short sessions of physical activity is as important as selecting the correct diet. The process of weight loss is a slow one, so owners who rely solely on reductions in their cat's bodyweight in order to gauge success will be rapidly de-motivated. Selecting behavioural targets can therefore be useful, as increases in levels of activity or interest in play are often noticed before the weight loss is registered.

5.4 Behaviour, Aging and Cognitive Dysfunction

Cognitive dysfunction is a medical condition, characterised by specific pathological changes within the central nervous system that result in well-documented, though sometimes subtle, behavioural changes. While there has been more research on canine than on feline cognitive dysfunction, recent work in the United States has confirmed that this age-related condition also exists in cats, and responds to a similar treatment regime. Of 152 feline patients over 11 years of age presented to a private American veterinary hospital, 43% exhibited signs consistent with cognitive dysfunction (Landsberg *et al.* 2003b). Some of these cats also had medical conditions that compounded their behaviour, but even when these individuals were removed from the analysis, 33% of cats over 11 years showed consistent signs. The percentage of affected individuals was higher in cats over 15 years (48%) than in those aged between 11 and 14 years (30%), and this older group of cats also exhibited more signs per cat. Since the long-term response to treatment is dependent on early detection of the condition, it is important that owners are aware of the potential significance of subtle changes in behaviour in their aging pets. Symptoms such as alterations in social relationships, sleep patterns and levels of activity, signs of anxiety and restlessness, spatial disorientation and confusion should all be monitored at regular intervals in the elderly feline. The provision of geriatric clinics within veterinary practices is to be encouraged.

5.5 The Effects of Stress on Disease

Associations between mental health, disease and welfare have been accepted in human medicine for some time, and it is now being increasingly recognised that such associations are also important in a veterinary context. The effects of stress on lower urinary tract function have been well established, but the effects of stress on susceptibility to, and spread of, infectious agents also have wide-ranging implications in cats (see Chapter 8). Since not all cats are obligate social animals, their housing in groups, whether in breeding establishments, catteries, shelters or other conditions, may have significant implications in terms of social compatibility and the potential for social stress (see Chapters 1, 2, and 7). The investigation of infectious disease outbreaks should therefore not only use traditional medical approaches, but also include a review of the affected cats' social situation and an assessment of potential stressors in the environment. Reorganization of social groupings and environmental changes, designed to maximise the availability of resources without the need for confrontation and conflict, should be an integral part of any treatment or management strategy.

6. CONCLUSIONS

In the early stages of the human-feline relationship, the cat's hunting ability was central to its co-existence with humans, who had a clear appreciation of its predatory skills. As companionship has surfaced as the most common reason for cat ownership in modern society, cats are increasingly expected to fulfil the expectations of their owners and the implications for feline welfare cannot be ignored. Misunderstandings of feline social behaviour have led to increasing numbers of multi-cat households, in which the stress of social incompatibility, and the pressure of inappropriate resource distribution, can result in a range of behavioural problems as well as the exacerbation of medical conditions. The need of cats to exhibit short bursts of high energy-consuming activity is viewed as problematic, and there is a mistaken belief that regular opportunities for play should be limited to the period of kittenhood. A misinterpretation of feline communication also contributes to the reporting of behavioural problems, so education of owners in terms of feline body language and communication signals is urgently needed.

Cats are highly adaptable, and in many ways are well suited to life as domestic pets. However, it is important that the development of their role as companion animals goes hand in hand with an understanding of their specific behavioural needs. An appreciation of the potential constraints in the domestic environment, and of methods to minimise their impact on feline welfare, is crucial.

7. REFERENCES

Appleby, D. L., Bradshaw, J. W. S. and Casey, R. A. (2001) The relationship between problematic canine aggression and avoidance behaviour, and experience in the first six months of life. *Veterinary Record* **150**, 434-439.

Association of Pet Behaviour Counsellors (2003) *Annual Review of Data*. Association of Pet Behaviour Counsellors, Worcester, UK.
http://www.apbc.org.uk/review_2003/report_03.htm

Beaver, B. V. (2003) *Feline Behavior: a guide for veterinarians.* 2nd edn., Saunders, St. Louis, Missouri.

Blue Cross (2003) *Parting with Pets Survey.* Blue Cross, Burford, Oxon, UK.

Borchelt, P. (1991) Cat elimination behavior problems. *Veterinary Clinics of North America: Small Animal Practice* **21**, 257-264.

Bradshaw, J.W.S. (1992) *The behaviour of the domestic cat.* CAB International, Wallingford, Oxon.

Bradshaw, J.W.S., Casey, R.A. and MacDonald, J.M. (2000) The occurrence of unwanted behaviour in the pet cat population. *Proceedings of the Companion Animal Behaviour Therapy Study Group Study Day*, Birmingham, pp. 41-42.

Bradshaw, J.W.S. and Hall, S.L. (1999) Affiliative behaviour of related and unrelated pairs of cats in catteries: a preliminary report. *Applied Animal Behaviour Science* **63**, 251-255.

Bradshaw, J.W.S and Lovett, R.E. (2003) Do domestic cats form hierarchies? *British Small Animal Veterinary Association Congress Scientific Proceedings,* Birmingham, p. 104.

Buffington, C.A.T. and Chew, D. J. (1998) Effects of diets on cats with non-obstructive lower urinary tract diseases: a review. *J. Animal Physiology and Animal Nutrition* **80**, 120-127.

Buffington, C.A.T., Chew, D. J. and Woodworth, B.E. (1999) Feline Interstitial Cystitis. *J. American Veterinary Medical Association* **215**, 682-687.

Buffington, C.A.T., Teng, B. and Somogyi, G.T. (2002) Norepinephrine content and adrenoceptor function in the bladder of cats with feline interstitial cystitis. *J. Urology* **167**, 1876-80.

Cameron, M.E., Casey, R.A., Bradshaw, J.W.S., Waran, N.K. and Gunn-Moore, D.A. (2004) A study of the environmental and behavioural factors that may be associated with feline idiopathic cystitis. *J. Small Animal Practice* **45**, 144-147.

Carlstead, K., Brown, J. L. and Strawn, W. (1993) Behavioural and physiological correlates of stress in laboratory cats. *Applied Animal Behaviour Science* **38**, 143-158.

Casey, R.A. (2001) A comparison of referred feline clinical behaviour cases with general population prevalence data. *British Small Animal Veterinary Association Congress Scientific Proceedings*, Birmingham, p. 529.

Cats Protection (2003) *Annual Review*. Cats Protection, Horsham, West Sussex, UK. http://www.cats.org.uk/html/pdf.php?file=annualreview2003.pdf

Collard, R.R. (1967) Fear of strangers and play behaviour in kittens with varied social experience. *Child Development* **38**, 877-891.

Corbett, L. K. (1979) Feeding ecology and social organisation of wildcats and domestic cats in Scotland. Ph.D. Thesis, University of Aberdeen, Scotland.

Crowell-Davis, S.L (2001a) Elimination behavior problems in cats. *Scientific Proceedings of the Annual Meeting of the American Animal Hospital Association* **68**, 34-37.

Crowell-Davis, S.L (2001b) Social organisation and communication in cats. *Scientific Proceedings of the Annual Meeting of the American Animal Hospital Association* **68**, 24-28.

Crowell-Davis, S.L., Barry, K. and Wolfe, R. (1997) Social behavior and aggressive problems of cats. *Veterinary Clinics of North America: Small Animal Practice* **27**, 549-568.

Dehasse, J., Heath, S., Muller, G. and Schubert, A. (2001) Feline Obesity: a behavioural approach. In Royal Canin's Guide to Feline Obesity, Royal Canin, Crown Pet Foods Ltd., Yeovil, Somerset, pp. 4-35.

Edney, A.T.B. (1998) Reasons for euthanasia of dogs and cats. *Veterinary Record* **143**, 114.

Feldman, H.N. (1994) Methods of scent marking in the domestic cat. *Canadian J. Zoology* **72**, 1093-1099.

Fitzgerald, B.M. and Turner, D.C. (2000) Hunting behaviour of domestic cats and their impact on prey populations. In Turner, D.C. and Bateson, P. (eds.). *The Domestic Cat: the biology of its behaviour*, 2nd edn., Cambridge University Press, Cambridge, pp. 152-175.

Gaskell, C.J. (2004) The lower urinary tract. In Chandler, G.A., Gaskell, C.J. and Gaskell, R.M. (eds.). *Feline Medicine and Therapeutics*, 3rd edn., Blackwell Publishing, Oxford, pp. 319-320.

Halip, J. (1994) Feline elimination problems. Presentation at the American Veterinary Medical Association Meeting, San Francisco, cited in Overall, K.L. (1997) *Clinical Behavioral Medicine for Small Animals*. Mosby, St Louis, Missouri.

Hall, S.L. and Bradshaw, J.W.S. (1998) The influence of hunger on object play by adult domestic cats. *Applied Animal Behaviour Science* **58**, 143-150.

Hand, M.S., Armstrong, P.J. and Allen, T.A. (1989) Obesity: occurrence, treatment, and prevention. *Veterinary Clinics of North America: Small Animal Practice* **19**, 447-474.

Heath, S.E. (2004) Common feline behavioural problems. In Chandler, G.A., Gaskell, C.J. and Gaskell, R.M. (eds.). *Feline Medicine and Therapeutics*, 3rd edn., Blackwell Publishing, Oxford, pp. 53-60.

Heath, S.E. (2002) Feline aggression. In Horwitz, D.F., Mills, D.S. and Heath, S. (eds.). *BSAVA Manual of Canine and Feline Behavioural Medicine*, British Small Animal Veterinary Association, Quedgeley, Gloucester, pp. 216-228.

Hetts, S. and Estep, D.Q. (1994) Behavior management: preventing elimination and destructive behavior problems. *Veterinary Forum* **11**, 60-61.

Horwitz, D.F. (2002) House soiling by cats. In Horwitz, D.F., Mills, D.S. and Heath, S. (eds.). *BSAVA Manual of Canine and Feline Behavioural Medicine*, British Small Animal Veterinary Association, Quedgeley, Gloucester, pp. 97-108.

Hunthausen, W. (2000) Evaluating a feline facial pheromone analogue to control urine spraying. *Veterinary Medicine* **95**, 151-155.

Hunthausen, W. and Seksel, K. (2002) Preventive behavioural medicine. In Horwitz, D.F., Mills, D.S. and Heath, S. (eds.). *BSAVA Manual of Canine and Feline Behavioural Medicine*, British Small Animal Veterinary Association, Quedgeley, Gloucester, pp. 49-60.

Karsh, E.B. (1983) The effects of early handling on the development of social bonds between cats and people. In Katcher, A.H. and Beck, A.M. (eds.). *New Perspectives on our Lives with Companion Animals*, University of Pennsylvania Press, Philadelphia, pp. 22-28.

Karsh, E.B. and Turner, D.C. (1988). The human-cat relationship. In Turner, D.C. and Bateson, P. (eds.). *The Domestic Cat: the biology of its behaviour*, 1st edn., Cambridge University Press, Cambridge, pp. 159-177.

Kitchener, A. (1991) *The natural history of the wild cats*. Christopher Helm Publishers Ltd, London, p. 70.

Landsberg, G., Hunthausen, W. and Ackerman, L. (2003a) *Handbook of Behaviour Problems of the Dog and Cat.* 2nd edn., Saunders, St. Louis, Missouri, pp. 220-222.

Landsberg, G., Moffatt, K. and Head, E. (2003b) Prevalence, clinical signs and treatment options for cognitive dysfunction in cats. *Proceedings of the 4th International Veterinary Behaviour Meeting,* Caloundra, Australia, pp. 77-83.

Markwell, P.J. and Butterwick, R.F. (1994) Obesity. In Wills, J.M. and Simpson, K.W. (eds.). *The Waltham book of clinical nutrition of the dog and cat,* Pergamon, Oxford, pp. 131-148.

Neville, P. and Bessant, C. (2000) *The Perfect Kitten: how to raise a problem free cat.* Hamlyn, London.

Overall, K. L. (1997) *Clinical Behavioral Medicine for Small Animals.* Mosby, St Louis, Missouri.

Pryor, P.A., Hart, B.L., Bain, M. J. and Cliff, K.D. (2001) Causes of urine marking in cats and effects of environmental management on the frequency of marking. *J. American Veterinary Medical Association* **219**, 1709-1713.

Schneirla, T.C., Rosenblatt, J.S. and Tobach, E. (1966) Maternal behaviour in the cat. In Rheingold, H.L. (ed.). *Maternal behaviour in mammals.* John Wiley, London, pp. 122-168.

Serpell, J.A. (2000) Domestication and history of the cat. In Turner, D.C. and Bateson, P. (eds.). *The Domestic Cat: the biology of its behaviour*, 2nd edn., Cambridge University Press, Cambridge, pp. 180-192.

Simpson, B.S. (1998) Feline House-soiling. Part II: Urine and faecal marking. *Compendium on Continuing Education for the Practicing Veterinarian* **20**, 331-339.

Spittler, H. (1978) cited in Fitzgerald, B.M. and Turner, D.C. (2000) Hunting behaviour of domestic cats and their impact on prey populations. In Turner, D.C. and Bateson, P. (eds.). *The Domestic Cat: the biology of its behaviour*, 2nd edn., Cambridge University Press, Cambridge, pp. 152-175.

Turner, D.C. and Meister, O. (1988) Hunting behaviour in the domestic cat. In Turner, D.C. and Bateson, P. (eds.). *The Domestic Cat: the biology of its behaviour*, 1st edn., Cambridge University Press, Cambridge, pp. 111-121.

Walker, A.D., Weaver, A.D. and Anderson, R.S. (1977) An epidemiological survey of the feline urological syndrome. *J. Small Animal Practice* **18**, 283-301.

Chapter 5

CAT OVERPOPULATION IN THE UNITED STATES

Philip H. Kass
Department of Population Health and Reproduction, School of Veterinary Medicine, University of California Davis, California 95616, USA

Abstract: The major cause of death of cats in the United States is neither infectious nor non-infectious disease, but the euthanasia of largely healthy cats in shelters due to the problem of overpopulation. This chapter explores overpopulation; while quantifying its magnitude is difficult, accurate measures are needed when assessing the effectiveness of interventions to control overpopulation, especially those conducted at the local level. There are many reasons why owners allow their cats to breed, despite the widespread availability of surgical sterilization and when early-age spaying and neutering are gaining favour in the veterinary medical community. In order to understand the main causes of overpopulation, these reasons, as well as the complex web of determinants of relinquishment of cats to shelters, are examined. Finally, this chapter evaluates the factors that influence whether relinquished cats are adopted or euthanized at animal shelters. Without concerted efforts to educate and inform the small minority of Americans who allow it to occur, cat overpopulation will continue to be a major welfare issue with no foreseeable resolution.

1. INTRODUCTION

Of all the issues affecting the welfare of companion animals in the United States, there can be none larger in scope, greater in magnitude, longer in duration and more worthy of disgrace than that of pet overpopulation. For decades, too many Americans have regarded companion animals as dispensable, and hence disposable. The origins of such indifference to the severance of a pet-owner relationship and the resultant animal suffering are difficult to fathom. A disheartening and sobering view is that this indifference starts in childhood, continues on into adulthood, and is perpetuated through example to future generations. Whatever the origins may be, the outcome is a society divided: a majority bereft at the immense scale of loss of animal life, and a minority who, despite the widespread availability of services to assist with the challenges of pet ownership, bears the primary responsibility for this unmitigated American tragedy.

This chapter provides an overview of how overpopulation manifests itself in the cat population in the United States. It begins by defining overpopulation and quantifying its magnitude, while acknowledging the imprecision of the estimates published thus far. An evaluation of uncontrolled breeding is presented, with an attempt to understand why it occurs and what could be done to prevent it. The relinquishment of cats to animal shelters plays a major role in the overpopulation problem, so a review of the research on relinquishment, as well as a review of factors affecting the adoption of cats from shelters, conclude the chapter.

2. DEFINITION AND ESTIMATION OF OVERPOPULATION

Measuring the magnitude of the cat overpopulation problem poses a formidable challenge. Indeed, there is not even a consensus about what overpopulation means; the euphemistic term "surplus" has been used as a presumably more socially acceptable substitute. Some argue that there is no excess of adoptable cats *per se*, but instead a lack of suitable and available homes (caused, for example, when residential covenants or lease constraints prohibit pet ownership). Others point to shelter statistics on the number of cats euthanized each year as *prima facie* evidence of overpopulation. In fact, the problem is more complex than either of these viewpoints suggest.

We can define cat overpopulation as the existence of cats that are at risk of euthanasia because they are both unwanted and not owned. This definition includes the most obvious manifestation of overpopulation: the dynamic population of cats in animal shelters. It is here that, for lack of an adoptive owner and adequate shelter resources, healthy, adoptable pets are euthanized, but overpopulation can be tangibly observed in other venues as well.

Veterinarians in private practice are frequently presented with pets that are unwanted for a multitude of reasons, varying from treatable behavioural problems to the sheer inconvenience of ownership. Many veterinarians decline to euthanize such pets on ethical grounds, though these beliefs are hardly universally felt among the veterinary community. With no legal requirement for reporting mortality and no universal standards for record keeping, the number of cats euthanized in hospital settings remains unknown. The presentation of cats to veterinary practices should not be construed, however, as an indication that all these former pets are not adoptable. Instead, the decision to euthanize these otherwise healthy pets arises from either the owner's decision not to allow the pet to be placed for adoption through a shelter, or the owner's belief – one not contravened by the veterinarian – that the cause leading to euthanasia in their healthy animal is neither treatable nor reversible. There are many causes of a breakdown in the human-animal relationship that owners, working together with veterinarians, could prevent or treat successfully but the owner, disregarding this, asks the veterinarian to bring their relationship with their pet to an end.

The issue of feral cats as contributors to overpopulation is controversial (see Chapter 6). Feral cats have vociferous detractors, who accuse them of serving as reservoirs for infectious diseases, including rabies, toxoplasmosis, feline leukaemia virus, and feline immunodeficiency virus. These diseases may be transmitted to other terrestrial and aquatic wildlife, and in some cases to humans. Feral cats may also be regarded as being responsible for a decline in certain native species, most notably songbirds. However, feral cats are not without their advocates. Many people "adopt" them, insofar as they provide them with food and, sometimes, additional care. Others believe that feral cats have essentially integrated themselves into the ecosystem where they live, occupying their own unique niche and becoming part of the local "balance of nature." This view erroneously assumes, however, that these cats are part of a stable bio-community, where animal and plant populations and species co-exist in a steady state. Usually, the integration of any new species, including cats, into a biotic community alters the structure in such a way as to destabilize it. A number of generations are required before stability develops, and displacement of native species is a possible, if not probable, result.

Feral cats may be born of other feral cats, or may be formerly owned cats that have adapted to living in the wild; they may live solitary lives or as part of a colony. Although there have been attempts to stop the reproduction of feral cat populations though population control measures, notably spay and neuter programs, there is little evidence to show that they have had meaningful impacts on the size of feral cat populations (see Chapter 6, where another view is presented). Many find the option of capturing, culling, and depopulation offensive; even when such cats are removed from the environment, others often arrive later to fill the niches left vacant.

Clearly, though, when feline overpopulation is mentioned attention turns to the municipal and private entities, known as animal shelters or animal control facilities, which harbor unowned or unwanted cats. Most of these facilities practice euthanasia, yet the magnitude of euthanasia defies determination for two principal reasons. Firstly, there is no complete comprehensive list of all shelters in the United States. Although local and state government funded shelters are easily located, many other shelters are operated privately, making their inventory difficult. Rowan (1992) contrasted the number of shelters estimated by the Humane Society of the United States (1,800) with the number of shelters estimated by the American Humane Association (3,000 to 5,000). Secondly, although the collection of shelter inflow and outflow statistics are usually mandatory in governmental shelters, and the data reported to the respective state agencies, the quality of such information is sometimes suspect. Record keeping may be far from precise, and in nongovernmental shelters such information gathering is not mandated nor is it always available for public inspection. Thus, shelter-based estimates of the number of cats euthanized annually are imprecise and probably an underestimate.

The number of cats reportedly euthanized each year has, predictably, varied widely. Between 5.7 and 9.5 million cats were estimated to be euthanized in animal shelters in 1990 (Kahler 1996). Another study, citing the American Humane Association in 1993, reported an annual figure of 5 to 7 million cats (Mahlow & Slater 1996). This differed from another report citing the American Humane Association, which claimed that 4 million cats were euthanized annually in shelters (Patronek et al. 1996). The lack of precision and consistency on a national level is less remarkable than the magnitude of even the lower estimates of cats euthanized annually.

In 1992 the State of California attempted to gather statewide information about animal shelter statistics, and with remarkable perseverance was able to achieve the participation of nearly all the governmentally-operated animal shelters in the state. This survey found that 484,173 cats were impounded at Californian shelters that year and 391,435 (81%) were ultimately euthanized; only 73,935 (15%) were adopted or reclaimed. Although these numbers are not precise – almost 4% of the impounded cats were unaccounted for – they provide a one-year, reasonably objective estimate of cat overpopulation, as measured by shelter euthanasia, in the state. California contains approximately 12% of the American population, so if the *per capita* euthanasia rate for California could be extrapolated to the entire country, we would estimate that approximately 3.26 million cats are euthanized each year. This number agrees with extrapolations of data from Massachusetts (3.18 million) and New Jersey and Washington combined (3.62 million) estimated by Rowan (1992), though is lower than the estimates provided by the American Humane Association. Whether these figures truly reflect the magnitude of feline overpopulation in American animal shelters rests on the assumption that they strictly refer to adoptable cats. In fact, this assumption is false. Shelters fulfill a critical need in American society for a less costly alternative to veterinary practices for euthanasia for humane considerations (Kass et al. 2001). Thus, the crude number of cats (and dogs) euthanized at shelters is not equivalent to the number of potentially adoptable cats (and dogs) euthanized at shelters. It is an overestimate: a substantial number of cats (and dogs) are relinquished to shelters in a non-adoptable condition for reasons that include serious and intractable behavioural problems, disease, and senescence. In a recently published study, 17% of all cats euthanized at 12 shelters in California, Colorado, Kentucky, New Jersey, New York, and Tennessee were euthanized for these reasons (Kass *et al.* 2001). The study found that the median age of cats presented to shelters specifically for euthanasia was 10 years, demonstrating that the pet-owner relationship had been, for many cats, a long and enduring one. Why, then, does an owner choose to have an older and/or ill pet cat euthanized at a shelter instead of a veterinary hospital? The decision is undoubtedly complex and includes considerations of prognosis, quality of life, anonymity, and economics. The latter reason is not trivial: costs of euthanasia at a veterinary hospital include those associated with an office visit, the euthanasia procedure, and one of several options for disposal of remains. In contrast, euthanasia at an animal shelter is usually performed without charge. These cats, which in most cases would not be suitable for adoption, should therefore be excluded from consideration when estimating the extent of pet overpopulation. Shelter statistics, as they are currently collected, present an overestimate of the number of adoptable pets euthanized simply for a lack of a deserving home.

While it makes little difference whether the estimates of cats unnecessarily euthanized are accurate on a national level, on a local level their accuracy becomes far more important. This is because it then becomes possible for municipal governments overseeing community shelters to assess the impact of intervention programs. The numbers of cats are smaller when measured in one shelter, or in a regional collection of shelters, than on a national scale, underscoring the need to be clear about the number euthanized for non-medical as opposed to medical reasons. This would, for example, allow a more precise determination of the effects of local spay/neuter and educational programs, designed to stem a shelter's inflow of unwanted yet adoptable pets as well as to enhance their outflow into appropriate homes. When the effects of an intervention over shorter periods of time are subtle, the need for accuracy becomes paramount. This would also encourage shelters to identify in their record keeping the cause of euthanasia for each relinquished animal. The collection of such information would go a long way to providing a better appreciation of the magnitude of feline overpopulation, at least insofar as it is evident in animal shelters. However, until the reporting of such information is legally mandated for all shelters practicing euthanasia, the magnitude can only be estimated.

3. THE PROBLEM OF UNCONTROLLED BREEDING

Many solutions to the problem of feline overpopulation have been suggested over the last several decades, but none have been more effective than the mass application of sterilization through the cooperation of municipalities, humane organizations, and public and private veterinary practices. The predominant method of controlling the size of the cat population in the United States is through surgical sterilization; alternative methods (contraceptive hormone therapy and vaccines) are not used or are experimental. Numerous surveys have shown that the proportion of pet cats that have been sterilized can exceed 80%, with allowances for geographic location, rural versus urban settings, owned versus relinquished cats, and licensed versus unlicensed cats (Nassar et al. 1983; Bloomberg 1996; Kass, unpublished data). The impact and importance of this degree of population control cannot be underestimated. Olson and Johnston (1993) demonstrated how two cats that give birth to eight kittens per year could, under reasonable assumptions, have about 175,000 descendants in seven years if breeding only occurs once per year, and over three quarter of a million cats if females are allowed to breed more often. The magnitude of cat overpopulation in the United States would reach unimaginable levels were it not for the widespread practice of surgical sterilization.

It is insufficient to sterilize pets, of course, if they are allowed to breed first. A segment of the American population believes that female cats need to experience parturition before they are sterilized. This was documented by Luke (1996), who found that slightly over 20% of cats gave birth to an average of 2.43 litters with an average size of 4.3 kittens, prior to being spayed. The reasons for this belief are unclear: among some individuals the tendency to anthropomorphize their parental instincts onto their pets may be the reason, while others may believe that parturition confers a tangible health benefit to the animal or that all animals have a right to reproduce. It is not obvious that such feelings could be dispelled even in the face of factual information, as they are as much a reflection of personal conviction and attitude as they are of personal knowledge, however misinformed.

There is also a widespread belief that cats need to reach sexual maturity before they can be sterilized. This may be, in part, a reflection of the formerly widespread advice of veterinarians that cats be at least six months old before being surgically altered. This advice is now being revised, as "early-age" spaying and neutering (pre-pubertal gonadectomy) becomes more widely accepted by the veterinary community. There was initial reticence about this practice because of the concern about stunting of normal growth patterns. There is little empirical basis to support this, although it is conceivable that long-term follow-up studies, if undertaken, may yet reveal unknown health effects (Olsen *et al.* 2001). However, given the tendency of some owners to procrastinate about sterilization until a cat breeds and produces a litter, it currently makes rational public policy to support the procedure of pre-pubertal gonadectomy in cats as young as seven weeks.

There are, of course, other reasons for the resistance to spaying and neutering, depending on the whims of individual cat owners, though none is more important than financial. Some owners, particularly those who have not established a close bond with their pet, feel the cost of surgical alteration, particularly in females, is prohibitive. Many communities have attempted to redress this problem by subsidizing low-cost spay and neuter programs, which have undoubtedly contributed to increasing the sterilization rate to over 80%. As these surgeries are often regarded as "loss-leaders" by veterinarians, and indeed cost substantially less than what the *de facto* equivalent of surgery (laparotomy and ovario-hysterectomy) in humans should cost, it is unlikely that they will become any less expensive on a widespread basis. Nevertheless, to the extent that communities can subsidize these operations even further and make them more widely available and affordable, the number of unsterilized cats will undoubtedly decrease.

It is unlikely that even greater access to the already widely available surgical sterilization could solve the overpopulation problem. First, while sterilization has helped to prevent a explosion in pet populations, the problem remains that a small but significant percentage of owners refuse to have their pets sterilized for a variety of reasons and convictions. Only slightly over half of cats relinquished to 12 shelters throughout the United States were sterilized (Salman *et al.* 1998). Second, many sterilized mature pets are still relinquished, evidence that overpopulation does not only arise from excess reproduction but also from a breakdown of an owner-pet relationship.

4. DETERMINANTS OF RELINQUISHMENT OF CATS TO ANIMAL SHELTERS

As overpopulation results from the breakdown of a pet's relationship with its owner, an understanding of the reasons for this breakdown is essential in order to develop methods to address it. What role did both the pets and their owners play in the relationship that ended, how did environment influence the outcome and what could have been done to prevent it? Few investigators have attempted to find answers to these questions. With the diversity of regions and peoples across the United States, generalization of a regional study to the entire country is tenuous, and the resources to conduct such work, whether measured by the availability of willing investigators or scarce finances, are difficult to come by.

Although there is much anecdotal information about factors associated with relinquishment, there is little peer-reviewed and published work on this subject. One of the earliest studies is that of Miller *et al.* (1996), which surveyed owners of dogs and cats relinquished to a humane society animal shelter in Hilliard, Ohio from October 1993 to January 1994; 74 cat owners ultimately completed the survey.

The authors asked these owners where they had initially acquired their cats. Almost half the cats (47%) came from private owners, with litters (17%), strays (14%), shelters (11%), and pet stores (11%) constituting the remainder. Surprisingly, approximately half of the owners had never intended to obtain the cat that they eventually relinquished; they had acquired it as a stray or as a gift. The most common age group of relinquished cats was 7 to 12 months (26%), followed by less than 6 months (24%), 19 to 24 months (14%), and 13 to 18 months (9%); only 18% of cats were older than three years. The principal reason for surrendering a cat to the shelter was that owners were moving from their residence (29%). Other common reasons included owner illness, particularly allergic reactions (15%); problematic cat behaviour, particularly fearfulness, scratching furniture, failure to eliminate in a litter box, and antipathy towards being handled; and being part of an unwanted litter (13%). Although the number of cats and owners in this study was comparatively small, this research paved the way for more studies.

Another study, also published in 1996, found somewhat different reasons for relinquishment (Luke 1996). Performed under the auspices of the Massachusetts Society for the Prevention of Cruelty to Animals, it examined the reasons for a cat being admitted to one of the organization's shelters between 1992 and 1994. Approximately 70,000 cats were admitted to these shelters during this period. The most common reason offered for surrendering a cat was ownership of a kitten that needed a home, reported by 24% of individuals, followed by unowned (stray) kittens needing a home, reported by 18% of individuals. The most common reasons for relinquishing adult cats were that the cat was a stray (17%), the owner was moving or had landlord problems (10%), cat behaviour problems (8%), the owner requesting euthanasia of the cat (7%), the owner no longer wanting the cat as a pet (7%), financial reasons (6%), and owner allergies (4%).

The most substantial body of work yet published on this subject in the United States arose in the late 1990s from the Regional Shelter Relinquishment Survey Study. This was a collaborative effort between the National Council on Pet Population Study and Policy (NCPPSP) and researchers at four American veterinary colleges or schools at Colorado

State University, Cornell University, the University of California, and the University of Tennessee (Salman *et al.* 1998). The strength of this study lay in its geographic diversity (it sampled 12 shelters in California, Colorado, Kentucky, New Jersey, New York, and Tennessee), its duration (one year), and its size (partial or complete information was obtained on almost 7,000 cats and dogs). The shelters were selected not only for their geographic diversity, but also to achieve a cross-section of shelters from rural, suburban, and urban environments. Shelters that specifically prohibited the euthanasia of animals ("no kill shelters") were excluded. The study sought to interview individuals contemporaneously with their relinquishment of a cat or dog at a shelter, using a questionnaire developed by individuals with expertise in shelter study and policy. Interviews were administered to 3,772 pet owners, 1,409 of whom relinquished cats or their litters.

The study determined that the most common classes of explanations for relinquishment of cats were issues related to human lifestyle (health and personal issues) (35%), issues related to human housing (26%), cat behavioural problems (not including aggression towards animals or people) (21%), the household animal population (15%), owner preparation for and expectation of pet ownership (15%), and request for euthanasia for reasons unrelated to old age and illness (12%). Other classes of explanations for relinquishment were found but were less frequently reported. Table 1 shows a detailed list of reasons for cat relinquishment reported by at least 10 cat owners (more than one reason could be provided for each cat).

Characteristics of cats surrendered to the shelters were evaluated. Only 8% of the cats relinquished were kittens, while 40% fell within the 5 month to 3 year age range, 23% were between 3 and 8 years, and 14% were older than 8 years (age was unreported in 16% of cats). A higher percentage of female cats (59%) were relinquished compared to male cats. Sterilization status was almost evenly split among cats (47% were intact, while 51% were neutered, and in 3% the status was unknown), and almost 93% were not purebred. Surprisingly, approximately 83% of the cats seldom or never went outdoors; these were true pets, as opposed to cats that owners rarely saw because they spent most of their lives outdoors. The cats were initially obtained from a wide variety of sources, the predominant ones being a friend who previously owned the cat (33%), a stray (23%), an animal shelter (14%), and offspring of an owner's pet (9%). The modal length of ownership of the relinquished cat was between 7 and 12 months (30%), followed by more than 5 years (19%), between 2 and 5 years (15%), and 1 to 2 years (16%); only 5% owned their cat less than 7 months (length of ownership was unreported in 15% of cats).

Table 1. Reasons given by owners for the relinquishment of their cats (n=1,409) to animal shelters: data from the Regional Shelter Relinquishment Survey Study. Percentages refer to the number of individual reasons for relinquishment divided by the total number of reasons for relinquishment (n=3,207) (Salman *et al.* 1998).

Reason for relinquishment	Number of reasons	Percent of reasons
Too many cats in home	348	11
Owner allergies	262	8
Moving	254	8
Cost	206	6
Landlord	198	6
No homes for litter	187	6
Soiling in house	163	5
Euthanasia for illness	159	5
Personal problems	119	4
Found cat	89	3
Euthanasia for old age	81	3
Illness	78	2
Inadequate facilities	68	2
Incompatible with other pets	66	2
Destruction inside the home	53	2
Aggression towards people	51	2
No time for cat	51	2
New baby in household	50	2
Responsibility	50	2
Abandoned	43	1
Aggressive to animals	42	1
Biting	38	1
Feral	37	1
Cat pregnant	34	1
Incompatible with children	33	1
Parent won't allow	33	1
Travel	32	1
Owner feared cat	32	1
Not friendly	28	1
Euthanasia for behaviour	25	1
Owner deceased	23	1
Too active	21	1
Owner pregnant	20	1
Attention	17	1
Disobedient	16	<1
Shedding fur	16	<1
Afraid	15	<1
Escapes	14	<1
Destroys outside	13	<1
Cat "jealousy"	11	<1
Divorce	11	<1
Wrong species	10	<1

Upon presentation to the shelters, 79% of owners requested that their cat be placed for adoption (although 96% described their cats as healthy), and 17% requested that their cat be euthanized. Only 47% of relinquishers had visited their veterinarian with their cat in the prior year, suggesting that veterinarians are under-used as a source of help to prevent pet relinquishment. Almost 24% of owners reported that their cats soiled in the home sometimes, mostly, or always, and a similar percentage gave the same frequency with respect to damaging the home. Other behaviours that relinquishing owners reported had occurred in their cats, within a year of their interview, sometimes, mostly, or always included hyperactivity (33%), noisiness (26%), fearfulness (44%), growling at people (13%), growling at animals (21%), attacking animals (13%), escaping (11%), biting a person (9%), and scratching a person (16%).

Information about the individuals relinquishing the cats was also available. Among those that provided information about their age, 7% were under the age of 21 years, 24% were between 21 and 30 years, 33% were between 31 and 40 years, 20% were between 41 and 50 years, and 16% were older than 50 years. Women brought cats into shelters more frequently than men (60% versus 40%). The percentage of relinquishers who had not attended high school was surprisingly high at 10%, and the percentage of those that only had a high school education was higher than expected at 35%. There was an inverse relationship between household income and frequency of relinquishment. A third of people relinquishing cats had a household income of under $20,000 per year, more than half of whom had incomes under $12,500 per year. In contrast, only 7% of relinquishers had an income of over $75,000 per year.

A sizeable number of people who brought their cats to shelters showed a diminished level of understanding of aspects of pet ownership, relative to what one might expect of owners in a successful pet-owner relationship. For example, 4% disagreed with the statement that animals need vaccinations or they may become ill; 58% agreed with the statement that animals misbehave out of spite; 14% disagreed with the statement that it is necessary for an owner to catch an animal misbehaving in order to correct it; 25% disagreed with the statement that cats can come into heat about twice per year; 29% agreed with the statement that rubbing a pet's nose in messes is effective discipline; 55% either agreed with the statement that animals are better off having a litter before being spayed, or were unsure if the statement was true; and 38% either agreed with the statement that cats don't mind sharing a house with other cats, or were unsure if the statement was true.

Additional work was then performed to develop a better understanding of the health and personal issues that together were the most common class of reasons for relinquishing a cat to a shelter (Scarlett *et al.* 1999). More than half the cat owners (59%) had owned their cat for less than one year. The most common reasons given for relinquishment were having family members with allergies to cats (18%), the owner experiencing personal problems (8%), the introduction of a new baby into the household (3%), and the owner not having time for the cat (3%). Among the cats relinquished because of allergies, 55% were relinquished within one year of joining the household, yet 15% of the cats had lived in the household for more than five years. Curiously, among the individuals relinquishing a cat for allergy reasons 11% nevertheless still owned at least one other cat. When this arose there were often other reasons given for relinquishment besides allergies, including housing issues, too many household pets, and non-aggressive behaviour problems. It is possible that allergies were not the true reason for relinquishment, but that some people found it a less socially stigmatizing excuse than other excuses. Certainly, the retention of other cats in the household when a person confesses to cat allergies leads one to be suspicious about the veracity of the claim.

Individuals with extremely low incomes (under $12,500 per year) seemed especially susceptible to relinquishing their cats for personal reasons: they constituted 44% of owners relinquishing for this reason. Also noteworthy was the finding that a relatively higher percentage of people citing personal problems as the reason for relinquishment requested euthanasia of their cats compared to people citing other reasons (8 to 11% versus less than 5%, respectively). People citing personal problems may have been under considerable stress when making the decision to have their pets euthanized, underscoring the need for counseling and intervention in certain situations. There were, interestingly, regional differences for at least one of the reasons given for this relinquishment class: allergies were reported as a reason more frequently in New Jersey and New York, and less frequently in California.

Of cats relinquished because of the owner moving residence, 52% were under the age of 3 years, and approximately 41% of cats were not sterilized (New et al. 1999). Almost half the cats had been owned for at least 2 years, and almost half of these had belonged to their owners for 5 years or more. Compared to the age distribution of the entire American population, owners in the 25 to 39 years age groups were over-represented among those relinquishing cats to shelters, and people in these younger age groups tend to be more mobile in their residences than older age groups. Seventy-one percent of the relinquishers identified their ethnicity as white. The authors concluded that rather than the act of moving alone, the interrelationships between age of relinquisher, duration of cat ownership, and possibly ethnicity may also play an important role. Although this is probably true, it is likewise correct that moving is associated with relinquishment for two primary reasons: either the destination is one at which cats are prohibited, or owners do not want to be bothered to take their cats with them when they move. Veterinarians are familiar with the latter situation, and some will decline to euthanize pets for this reason. The former reason is more problematic, and symptomatic of a larger issue in American society: landlord prohibitions of pet ownership can supersede a renter's right to enjoy a pet-owner relationship. Resistance of landlords to pet ownership remains difficult to address, despite the availability of compromises such as damage deposits and constraining the size of the pet to be appropriate to the size of the residence. Although some communities have banned prohibitions on pet ownership as discriminatory, it is all too common to find that people who rent their residences are faced with the choice between keeping their economical housing or losing their pet. Another way that moving influences relinquishment is when young adults who have a cat (or dog) leave the parental home, but cannot find a rental that will allow pets. The parents of these young adults are then left, sometimes unwillingly, with the task of caring for these pets, and eventually relinquish them to shelters.

Another study done shortly before the Regional Shelter Survey Study is that of Patronek *et al.* (1996). It was on a much smaller scale: relinquished cats came from a single shelter in Indiana over a nine-month period. It differed from the larger study in one important point: it included a comparison group of current cat-owning households, thus assuming the features of a population-based case-control study. This made possible the calculation of quantitative relative effect measures, something the Regional Shelter Survey Study alone could not do. A total of 218 cats was relinquished to the study shelter, and these cats (and their owners) were compared with their respective counterparts in the comparison group (n=459).

The authors found that age was again a determinant of relinquishment: cats under 6 months were approximately 14 times more likely to be relinquished as cats older than 5 years. Although the study did not assess age effects after 5 years, there was a clear trend towards diminished risk of relinquishment as cats aged up to 5 years. Cats that owners obtained without any payment to a previous owner were more frequently relinquished than cats purchased or adopted. Curiously, cats acquired as strays were at lower risk of relinquishment than cats acquired from a private owner or cat breeder.

A number of ownership characteristics were likewise predictive of relinquishment. Some owners reported confining their cats for parts of the day to their basements or garages; these cats were more likely to be relinquished compared to cats allowed to wander around the house unrestrained. Likewise, cats that were allowed to roam outdoors had an increased risk of relinquishment. When an entire family provided for the cat, instead of just a female adult, the probability of the family surrendering the cat diminished. Owners that ultimately relinquished their cats sought behavioural advice, mostly from veterinarians, at a higher frequency than owners in the comparison group. Yet predictably the relinquishing owners were more likely to find the advice impractical or unhelpful, or not act on it at all. Failure to ever take the cat to a veterinarian increased the risk of relinquishment approximately four-fold. However, the number of annual veterinary visits was not associated with risk of relinquishment after controlling for age, sterilization, declaw status, duration of ownership, and outdoor access.

Behavioural problems figured prominently as reasons for relinquishment. Increasing frequency of inappropriate elimination, inappropriate scratching, and aggression toward people were all associated with an increased probability of a cat being surrendered to a shelter. Prior ownership experience of another cat had a protective effect against relinquishment, possibly in part because of more realistic expectations of the role of the cat as a member of a household.

As found in the Regional Shelter Survey Study, household income was an important predictor of the likelihood of relinquishment. The investigators determined that households with incomes under $20,000 were approximately four times more likely to relinquish a cat than households with incomes greater then $75,000. Furthermore, people renting an apartment were almost three times more likely to take a pet cat to a shelter than people who lived in a single-family dwelling. Similarly, not owning a home increased the likelihood of a cat being relinquished more than twice compared to owning a home, and the length of time living in a home was strongly inversely proportional to the probability of relinquishment. Finally, education again was predictive of risk of relinquishment: among individuals who at least completed high school, the risk of relinquishment was lowest among individuals with postgraduate degrees, followed closely by college graduates. The educational class with the highest risk was individuals with a post-high school vocational education, while paradoxically the lack of a high school degree did not seem to have any effect on relinquishment probability compared to people with postgraduate education. The authors concluded that the most significant factors influencing the surrender of cats to the shelter were having specific expectations of the cat as a household member, owning a non-sterilized cat, not reading any materials about cat behaviour, the cat showing inappropriate elimination behaviours, allowing the cat access to the outdoors, and unrealistic expectations about the care needed for a pet cat.

The authors of the Regional Shelter Relinquishment Survey later published additional work to further the inferences that could be made from their original research (New et al. 2000). The newer efforts included a comparison group of households that owned at least one cat or dog in 1996, approximately half of which had pets leave the household during that same year (the most common reasons for cats leaving were that the cat died (33%), was euthanized (19%), was given away (13%), disappeared (17%), was relinquished to an animal shelter (4%), or was sold (0.4%)). The design and analysis took on the features of a case-control study similar to that performed by Patronek *et al.* (1996), but differed in that the comparison group was not a sample from the source population of animals relinquished. The latter came from 12 shelters in six states (Salman *et al.* 1998), while the comparison group was a sample of households from throughout the United States. However, unless households in successful pet-owner relationships in the regions studied in the Regional Shelter Relinquishment Survey are exchangeable with similar households from regions throughout the United States with respect to responses to the survey metric, the results from the newer study are not, strictly speaking, interpretable as effect measures.

In a comparison of ages between the relinquished and non-relinquished groups, kittens were clearly at the highest risk of being relinquished, with the risk steadily decreasing as age increased to 3 years. Cats in the 10 to 14 year age range had the lowest risk of any age class, and risk increased again at age 15 years or later (presumably because many of these cats were not relinquished for adoption, but rather for euthanasia due to age-related health problems). Purebred cats had a diminished risk of more than 50% compared to mixed breed cats, which was the predominant breed class in the study. Having a cat sterilized apparently reduced the risk of surrender: the frequency of intact male cats was almost three times greater in surrendered compared to non-surrendered cats, and in female cats was almost four times greater. Four sources of cats appeared to be associated with an elevated risk of relinquishment: friends, pet shops, breeders, and shelters; the lowest risk was seen in offspring of the owners' own cats. Length of cat ownership was also clearly a determinant of relinquishment: compared to ownership of 15 years of more, ownership of less than 3 months was almost 30 times more common in shelter compared to owned cats, approximately nine times more common for 3 to 5 month ownership, and two to three times more common for 6 months to under 2 years of ownership. Not unexpectedly given the above findings for age, ownership for 10 to 14 years had the lowest risk of relinquishment of any length of ownership class, including ownership of 15 years or more. Although the study found that cats purchased for more than $100 appeared to be less likely to be relinquished, the number of cats that cost this much (n=15) was too small to reach meaningful conclusions about the trend in cost versus probability of surrender. Interestingly, a history of biting a person was no more common in the relinquished cats than in the currently owned cats.

Questions about the frequency of undesirable cat behaviour often revealed wide disparities between owners who surrendered their cats to shelters compared to owners in successful cat-owner relationships. The frequency of cats soiling mostly, almost always, or always in the house was reported twice as frequently in surrendering compared with non-surrendering households. Virtually the same association was observed for cats mostly, almost always, or always damaging items either inside or outside the home. Perception that hyperactivity in the pet cat as almost always or always present was found over four times more frequently, and as mostly present over two times more frequently, in surrendering versus non-surrendering households. In contrast, there were few meaningful differences in frequencies reported between the two household types with respect to the following behaviours: vocalization; fearfulness of people, animals, noises, or objects; growling, hissing, snapping, or attempting to bite people or other animals; and attacking or initiating a fight with another animal. Curiously, when asked how frequently the cat escaped from the house or yard, owners relinquishing their cat reported this behaviour at least some of the time only approximately half as often as control households. Although the reason for this finding was not explored, it seems plausible that it may be more of a reflection of surrendering households being less aware of their cat's activities, than of a diminished frequency of this undesirable behavioural trait in these households.

When owners were posed questions designed to elicit their understanding of cat behaviour, significant differences were noted between household types. For example, when asked if cats did not object to the presence of other cats in the home, 14% of surrendering households did not know an answer, compared to 21% of non-surrendering households. Another question asked if cats pounced, scratched, or bit as part of playing, and while 93% of households in successful cat-owner relationships responded affirmatively, only 86% of relinquishing households agreed. Most telling was the disparity in responses to a statement that a female cat will be "better off" if she has one litter before being "fixed." Among the non-relinquishing households only 13% agreed with the statement, in contrast with 21% of relinquishing households.

As noted in other studies, young adults were more likely to surrender their cats to shelters than other age classes. The highest risk age group found in this study was 20 to 24 years, followed by under 20 years, 25 to 29 years, and 30 to 34 years. Few differences in age classes were noted after the age of 35 years. As expected, education was likewise a predictor of relinquishment. Compared to people with some education following high school, the increase in frequency of relinquishment was 70% in males with, at most, a high school education, and 90% in females with, at most, a high school education.

It is hazardous to make overly broad generalizations about cat relinquishment, when every case is unique, difficult, and complex (DiGiacomo et al. 1998), and particularly in the United States where there may be considerable heterogeneity across regions. Nevertheless, the studies cited above share several common findings. The first is that there is an element of inexperience and ignorance that is common among cat owners and can contribute to the failure of the cat-owner relationship. Age, education, income, and lack of access to behavioural and veterinary care are certainly important, but it is possible that a lack of appropriate role models, including parental role models, may also play a substantial part. Secondly, there are certain problematic behaviours in cats that lead owners to frustration and sometimes the decision to turn over cats to shelters. Although the single and joint effects of these factors should not be minimized - indeed, it is unlikely that all relinquishment of healthy adoptable cats can be eliminated - it is realistic to believe that with appropriate interventions, the number of relinquished animals can be meaningfully decreased. Such interventions are neither trivial nor inexpensive, and must target individuals both at the stage when they obtain their pet, as well as when problems arise during the course of ownership. Given the heterogeneity of the American population, individual communities will need to develop their own individual strategies.

5. DETERMINANTS OF ADOPTION OF CATS FROM ANIMAL SHELTERS

Cat overpopulation can be visualized as a prevalence pool of adoptable and unwanted cats that has both inflow and outflow components. To address overpopulation it is necessary to study not only the inflow component, which includes factors such as overbreeding, failure to sterilize pets and the breakdown of the pet-owner relationship, as discussed previously, but also the outflow component, which consists primarily of the adoption of cats from shelters. Yet, while little has been scientifically documented about the circumstances surrounding cat relinquishment to shelters, there are even fewer studies about the factors that affect the likelihood of cats being adopted from shelters.

Determinants of adoption of dogs from animal shelters have been studied in California within the last 10 years (Barnes 1995; Clevenger & Kass 2003). However, only one study has evaluated the factors that have distinguished adoptions from euthanasias in cats (Lepper et al. 2002). It was conducted at the Sacramento County Animal Shelter between September 1994 and May 1995 and included 3,301 cats that were offered to the public for adoption, of which only 670 (20%) were ultimately adopted; the remaining cats were euthanized. It is well known among shelters throughout the United States that the most adoptable cats are kittens, and this was borne out by this study. Cats under 1 year of age were adopted approximately four times more often than cats between 1 and 2 years of age, approximately five times more often than cats 3 to 5 years of age, and approximately nineteen times more often than cats over 5 years old.

Predictably, gender and sterilization status influenced the decision of owners to adopt cats (Figure 1). Sterilized cats were preferred over sexually intact ones: spayed females were over four times more likely to be adopted as intact females, and neutered males were almost six times more likely to be adopted as intact males. Neutered males were preferred by a factor of 1.56 over spayed females, and intact males were preferred by a factor of 1.17 over intact females.

Figure 1. Factors affecting the likelihood of a cat being adopted include its age, sex, sterilization status, fur colour, breed, and the reason for relinquishment of the cat to the shelter. (Courtesy of Irene Rochlitz).

The cat's fur colour and breed also appeared to influence the choice of a pet. The prototypical "tabby" coloration was used as a reference against which other coat colours were compared. White cats were the favourite colour, followed by colourpoint and gray coats. In contrast, predominantly brown or black cats were almost half as likely as tabby cats to be adopted. The majority of cats in the United States are collectively referred to as "domestic short hair." When this breed group was compared against all other breeds, the authors found a preference for Persians and other rare purebred cats, which were almost twice as likely to be adopted. In contrast, cats loosely classified as "domestic medium hair" and "domestic long hair", as well as Siamese cats, had roughly the same likelihood of adoption as the "domestic short hair" cats.

Sixty percent of the cats placed for adoption were considered to be "stray" upon relinquishment. These included unowned cats brought to shelters by individuals and cats picked up by animal control officers. Many other reasons for relinquishment other than stray were given, and as expected some cats (including feral and injured) were not well represented in the group of adoptable cats. In general, the more common reasons for relinquishment, compared to being presented as a stray, reduced the likelihood of adoption. These reasons included the cat having behavioural problems, being classified as neonatal (under eight weeks old), and being described as ill or geriatric.

Not all adoptions of cats from animal shelters result in long-term successful relationships between owners and their pets. Work by Kidd *et al.* (1992a, 1992b) revealed some factors that distinguish successful from unsuccessful adoptions. In the first study, the authors performed a six-month follow-up of 161 owners who adopted cats from a shelter. They found that women adopted cats at a significantly higher frequency than men (53% versus 35%). A higher percentage of men than women, and of parents than non-parents, did not keep their cat for longer than six months. Of those individuals who rejected their pets, 62% were first-time adopters while 38% had previously owned pets; this difference was significant. Those rejecting their pets were on average younger than those who retained their pets beyond 6 months. In the second study, the authors observed that people adopting directly from veterinarians were more likely to retain their cats than those adopting from a shelter. The average age of the rejecting owners was approximately 8 years lower than non-rejecting owners, and the average age of the veterinary clients was approximately 7 years older than the shelter adopters. The authors also found that, in general, when veterinary owners relinquished their cats they had managed to keep them for an average of 6 months following acquisition, in contrast with the shelter adopters, most of whom surrendered their cat within 2 months. The conclusion of the investigators was that by providing information about pet care and training, imparting upon owners more realistic expectations about normal pet behaviour and how to mitigate undesirable behaviour, improving an understanding of what is involved with pet ownership, and impressing upon owners the need to recognize their commitment to their pet, it is possible to reduce the number of newly adopted pets that are ultimately rejected.

6. CONCLUSIONS

Solving the problem of cat overpopulation in the United States remains a difficult, complex and elusive task, and the likelihood of having a substantial impact in the near term is poor. Although the majority of owners responsibly prevent their pets from breeding, there remain a sizable number of people who either allow their pets to have offspring before sterilizing them, or refuse to sterilize them for reasons that are usually either economic, based on a conscious desire to breed and raise offspring, or belie a basic misunderstanding of animal biology.

The decision to ultimately relinquish pets is multi-factorial and complex, but fundamentally reflects a pet-owner relationship that is lacking in profound emotional attachment. Compared with kittens, those animals that are victims of the breakdown of their bonding with their owners are among the least likely to be adopted from shelters. The main causes of relinquishment, discussed in this chapter, defy simple interventions and addressing them may require strategies that take into account issues particular to a specific region of the United States. A fundamental re-education of the American public is also required, teaching the younger and older generations the importance and value of their pet's life, so that companion animals will no longer be regarded as dispensable and disposable.

7. REFERENCES

Barnes, D.D. (1995) Retrospective cohort study of factors affecting time to adoption of dogs in a humane society. M.Sc. Thesis, University of California, Davis.

Bloomberg, M.S. (1996) Surgical neutering and nonsurgical alternatives. *J. American Veterinary Medical Association* **208**, 517-519.

Clevenger, E.J. and Kass, P.H. (2003) Determinants of adoption and euthanasia of shelter dogs spayed or neutered in the University of California Veterinary Student Surgery Program compared to other shelter dogs. *J. Veterinary Medical Education* **30**, 372-378.

DiGiacomo, N., Arluke, A. and Patronek, G. (1998) Surrendering pets to shelters: the relinquisher's perspective. *Anthrozoös* **11**, 41-51.

Kahler, S.C. (1996) Welfare of cats depends on humankind. *J. American Veterinary Medical Association* **208**, 169-171.

Kass, P.H., New, J.C. Jr., Scarlett, J.M. and Salman, M.D. (2001) Understanding animal companion surplus in the United States: Relinquishment of nonadoptables to animal shelters for euthanasia. *J. Applied Animal Welfare Science* **4**, 237-248.

Kidd, A.H., Kidd, R.M. and George, C.C. (1992a) Successful and unsuccessful pet adoptions. *Psychological Reports* **70**, 547-561.

Kidd, A.H., Kidd, R.M. and George, C.C. (1992b) Veterinarians and successful pet adoptions. *Psychological Reports* **71**, 551-557.

Lepper, M., Kass, P.H. and Hart, L.A. (2002) Prediction of adoption versus euthanasia among dogs and cats in a California animal shelter. *J. Applied Animal Welfare Science* **5**, 29-42.

Luke, C. (1996) Animal shelter issues. *J. American Veterinary Medical Association* **208**, 524-527.

Mahlow, J.C. and Slater, M.R. (1996) Current issues in the control of stray and feral cats. *J. American Veterinary Medical Association* **209**, 2016-2020.

Miller, D.D., Staats, S.R., Partlo, C. and Rada, K. (1996) Factors associated with the decision to surrender a pet to an animal shelter. *J. American Veterinary Medical Association* **209**, 738-742.

Nassar, R., Mosier, J.E. and Williams, L.W. (1983) Study of the feline and canine populations in the Greater Las Vegas area. *American J. Veterinary Research* **45**, 282-287.

New, J.C. Jr., Salman, M.D., Scarlett, J.M., Kass, P.H., Vaughn, J.A., Scherr, S. and Kelch, W.J. (1999) Moving: characteristics of dogs and cats and those relinquishing them to 12 U.S. animal shelters. *J. Applied Animal Welfare Science* **2**, 83-96.

New, J.C. Jr., Salman, M.D., King, M., Scarlett, J.M., Kass, P.H. and Hutchison, J.M. (2000) Characteristics of shelter-relinquished animals and their owners compared with animals and their owners in U.S. pet-owning households. *J. Applied Animal Welfare Science* **3**, 179-202.

Olson, P.N. and Johnston, S.D. (1993) New developments in small animal population control. *J. American Veterinary Medical Association* **202**, 904-909.

Olson, P.N., Root Kustritz, M.V. and Johnston, S.D. (2001) Early-age neutering of dogs and cats in the United States (a review). *J. Reproduction and Fertility Supplement* **57**, 223-232.

Patronek, G.J., Glickman, L.T., Beck, A.M., McCabe, G.P. and Ecker, C. (1996) Risk factors for relinquishment of cats to an animal shelter. *J. American Veterinary Medical Association* **209**, 582-588.

Rowan, A.N. (1992) Shelters and pet overpopulation: a statistical black hole. *Anthrozoös* **5**, 140-143.

Salman, M.D., New, J.G. Jr., Scarlett, J.M. and Kass, P.H. (1998) Human and animal factors related to the relinquishment of dogs and cats in 12 selected animal shelters in the United States. *J. Applied Animal Welfare Science* **1**, 207-226.

Scarlett, J.M., Salman, M.D., New, J.G. Jr. and Kass, P.H. (1999) Reasons for relinquishment of companion animals in U.S. animal shelters: selected health and personal issues. *J. Applied Animal Welfare Science* **2**, 41-57.

Chapter 6

THE WELFARE OF FERAL CATS

Margaret R. Slater
Department of Veterinary Anatomy and Public Health, College of Veterinary Medicine, Texas A&M University, College Station, TX 77843-4458, USA

Abstract: The control of feral cats is a controversial issue in many countries, due to the differences in the way humans perceive cats in general and feral cats in particular. As cats spread into a wide range of habitats, there are concerns regarding the best methods to control their numbers. Predation on wildlife, public health and zoonotic diseases, as well as the welfare of the cats themselves, are issues that drive the need to control the feral cat population. Killing the cats, or letting nature take its course, were the usual historical approaches but in recent years non-lethal methods have been espoused as being more humane and effective. Efforts have been made to improve the welfare of feral cat populations through sterilization, the control of infectious disease and ensuring that they are adequately cared for. A combination of approaches are necessary to decrease feral cat numbers, to prevent influx of owned cats into the population, and to manage established feral cat colonies successfully.

1. INTRODUCTION

For over four thousand years, cats have closely accompanied the development of human society, both as real and as symbolic creatures. Often associated with evil, witchcraft, devil worship or simply bad luck, they have at times been used as scapegoats for natural disasters or personal misfortunes (Tabor 1983; Serpell 2000). While some of these negative stereotypes persist to the present day, ever since the 19th century there has been a rapid evolution towards a far more favourable perception of cats.

In many countries the welfare of all cats, and in particular feral cats, has become a focus of public concern. Feral cats are likely to be found wherever humans have traveled, either as escapees from domestication or as deliberately introduced controllers of rodents or other pests (Figure 1). The interest in feral cats may focus on animal control, especially in countries where the free-roaming dog problem no longer is a major concern, or on issues such as predation, public health or the well-being of the cats themselves.

Figure 1. Feral cats are often found in public locations such as restaurants.

Feral cats are still viewed by many as creatures living on the borders of civilized communities. This view reinforces the peripheral status of cats and emphasizes their wild or natural propensities. There are also those who argue that feral cats do not belong in the wild, because they are introduced predators of more valued species. I hope to throw some light on the discussion about where feral cats belong, and how to deal with them, by examining selected English language publications, particularly those from the past fifteen years. I have used the scientific literature when it is available but for some types of information, lay publications and personal communication are the only available sources.

2. DEFINITIONS OF FERAL CATS

A wide variety of terms are used to describe feral cats, such as free-roaming cats, barn cats, stray cats, etc. This can make comparisons between studies difficult. For example, in one study stray cats were defined as those taken from dumps, residential or industrial areas and ferals as those remote from these locations (Read & Bowen 2001). Other definitions have related to the reliance on humans for food and shelter (strays) or independence of humans (ferals) (Hugh-Jones *et al.* 1995; Dickman 1996b).

I have not defined feral cats in the usual biological sense, which views them as having "escaped" domestication and gone wild, or as having populations which reproduce in the wild. My definition of a feral cat is a pragmatic one, based on the status of an individual cat at a particular point in time. A feral cat is one that cannot be handled and is not suitable for placement into a typical pet home, that is, a cat that is unsocialized.

Socialization is defined as the process by which an animal develops appropriate social behaviour toward conspecifics (Turner 2000). However, the term is commonly used to describe the relationship between cats and humans, in the context of "socialization to humans". I use the term socialized rather than tame, as it is a more accurate description of those cats that are adoptable. The socialized or unsocialized (feral) status must be determined, recognizing that there is considerable variability among cats which may be modified by the situation and change with time. The experience, knowledge and type of interaction of the observer may have a great influence over the gestalt assessment of each cat's sociability. Many factors have been shown to affect the socialization of cats, and are discussed in Chapter 3.

2.1 Socialization Status, Ownership Level and Confinement Level

In general, the sociability of a cat relates to its comfort when handled by a person. The sociability index is a spectrum, ranging from cats that are completely unfamiliar with humans, are terrified of them and cannot be handled (feral cats), through cats that have some limited interaction with familiar caretakers, to cats that are very social and friendly. A stray cat is an owned cat that is lost, or has been abandoned by an owner. Stray cats are usually considered to be socialized since they were in a household in the recent past.

Ownership level refers to the degree of care and commitment provided by people towards cats. At one end of the spectrum are cats considered by their owners as members of the family, and whose social, environmental and health needs are provided for. At the other end are cats that are not cared for by humans, and in between are cats that receive some level of care from a specific person or household, or receive regular but limited care by caretakers.

Another concept is the confinement level of the cat. Confinement ranges from completely indoor cats, to cats confined to the owners' property, to cats that are allowed to roam some or all of the time. Generally feral cats are not confined. Cats that roam freely, at least for part of the time, are those that usually cause problems and concerns.

2.2 Other Descriptors Used for Cats

Terms such as barn cat, alley cat, doorstep cat, etc. are used to refer to the locations of the cats. These terms should only be used to describe the location of the cats and not to imply their sociability or ownership status.

A colony is a group of three or more sexually mature cats living and feeding in close proximity. A queen and her nursing kittens are not a colony as the kittens are still dependent on the mother and immature. This situation has been described as a "proto-colony" since, in time, the kittens and queen will likely become a colony. A managed colony is a colony that is controlled by a trap, neuter and return approach (see section 5.3).

3. SOURCES OF FERAL CATS

Feral cats may be the offspring of existing feral cats, lost or abandoned cats that have become unsocialized or the offspring of owned, intact cats allowed outside. The relative importance of each source will vary widely from location to location and has rarely been studied.

Data on cat ownership in different countries are presented in Chapter 3. Data on the proportion of owned cats that are allowed to roam are not available, and probably vary widely between countries. There may be many stray cats that could potentially become feral. Studies in the United States found that about 22% of owned cats were acquired as strays (New, Jr. et al. 2000). Another study of a single community found that 25% of owned pets were former strays and 63% of cats entering a shelter were impounded strays (Patronek et al. 1997). Litters from owned cats are another potential source of feral cats. One study in Massachusetts reported that over 90% of all cats (male and female) were sterilized. In spite of this high sterilization rate 15% of currently sterilized female cats had previously had litters, with a similar number of total litters per female for intact and sterilized cats (Manning & Rowan 1998).

4. NUMBERS OF FREE-ROAMING AND FERAL CATS

Studies from the west, south and northeast United States indicate that between 9 and 22% of households feed free-roaming cats that they do not own (Haspel & Calhoon 1993; Johnson *et al.* 1993; Johnson & Lewellen 1995; Luke 1996; Levy *et al.* 2003b). It has been suggested that in the United States the number of free-roaming cats equals the number of owned cats (Holton & Manzoor 1993), but others believe that the number ranges from 25 to 60 million (Patronek 1998). In warmer climates there may be larger numbers of free-roaming unowned cats, since females are able to produce two to three litters in a prolonged warm season and mild winters will result in lower mortality. The number of free-roaming cats will also depend upon the popularity of cats as pets, the beliefs of the owners regarding cats' need to go outdoors, the sterilization rate, the availability of food and shelter and the existence of other predators (see Chapters 3 and 5).

The proportion of free-roaming cats that are feral will vary with the location. Anecdotally, between 10 and 50% of the total cat population taken in by animal control facilities in the United States are feral. Management programs using trap, neuter and return, often find that between 50 and 90% of cats in colonies are feral. Based on my experience, I estimate that the number of feral cats in the United States is about one third to one half the number of owned cats.

Cats have a social structure that lies between the larger pack-hunting carnivores like lions and the solitary territorial leopard and wild cat (Fitzgerald & Karl 1986) (see Chapter 1). Free-roaming cat populations appear to be controlled by the resource dispersion hypothesis, suggesting that availability of food is the primary limiting resource for female cats and drives their dispersion (Macdonald *et al.* 2000). Food-driven dispersion may itself be mediated by other resources such as shelter or resting places, and by competition with other animals (Calhoon & Haspel 1989; Liberg *et al.* 2000; Macdonald *et al.* 2000). Thus, the presence of a localized, stable and large food source appears to be the primary reason for group living in domestic cats (Smith & Shane 1986; Liberg *et al.* 2000). For male cats, another limiting resource for group living will be access to females (Macdonald *et al.* 2000).

5. METHODS FOR CONTROLLING FERAL CAT POPULATIONS

The "wait and see" or "do nothing" approach to controlling cat populations has been used historically, and is still applied in some locations. The hope is that "nature will take its course" and cats will be killed or move away. In reality, doing nothing is a poor choice for the cats and is not a solution to the problem. Therefore, methods for dealing with populations of free-roaming and feral cats have been developed, and can be divided into three main approaches. The first is to kill cats on site, the second is to trap and remove cats for euthanasia or relocation, and the third is to trap, neuter and return cats to the original location.

Human perceptions of cats influence the selection of methods used to control them. A review of the factors that influence the way humans perceive cats are discussed in Chapter 3.

5.1   Kill on Site

Cats have been accidentally or deliberately introduced to a broad range of locations, including islands (Courchamp & Sugihara 1999). Since cats are very adaptable, they have survived and reproduced, accommodating to different food sources. They are also fecund, giving birth to one to three litters per year of two to six kittens. Since cats are sexually mature at five or six months of age, this can result in a substantial number of cats in a short period, even with very high mortality rates. Moreover, if they die from disease or human intervention, other cats move in to take advantage of the newly available space and food supplies (Tabor 1983). This is especially the case in locations that are not geographically isolated. These facts suggest that wholesale slaughtering is not a practical solution to permanently eliminating a colony.

Methods of killing cats on location are generally not popular with the public. They are usually used in locations without human habitation, and are often chosen by local governments because they are perceived to be permanent, relatively inexpensive solutions to feral cat problems. Poisons are not specific and may endanger other animals and humans as well as causing a painful death for the cats, so they should only be used in very specific settings (Dowding *et al.* 1999).

Increasingly, the public views cats as domestic animals for whom it has a responsibility, and does not accept the killing of cats as a solution to a problem that, in many instances, is due to people introducing cats to the location in the first place. This view arises from the change in the perception of non-human animals from property, incapable of feelings or thoughts, to animals as companions that experience pain, hunger and other emotions. An example of this change and how it affects animals occurred in April 2003, in the cities of Mataro and Barcelona, Spain. These cities prohibited shelters from destroying stray cats and dogs that were not severely ill, injured or dangerous (www.aldf.org). This change appears to have been due to actions by several animal welfare and rescue organizations, one of which recently took over the government shelter in Barcelona. In Italy, a 1991 law prohibited the abuse or removal of feral cats from their colonies and made provision for the Public Veterinary Services to sterilize the cats (Natoli *et al.*

1999). The trap, neuter and return method (see section 5.3) has become widespread in Italy. More recently, the cats of Rome were given the status of "patrimonio bioculturale" that is, that they are a bio-cultural heritage (www.romancats.de/romancats/news/article.php?Id=100) (Figure 2).

Figure 2. While the feral cat colonies of Rome are well known, feral cats can be found throughout Italy.

5.1.1 Efficacy of Kill on Site Methods

When determining the best option for controlling feral cat populations in natural settings, particularly islands, well-designed studies are required to provide reliable data on the effects of feral cats on wildlife and to devise appropriate management programs (Dickman 1996b). All other non-native species must also be monitored, and native species at highest risk should be identified. Other factors that are likely to interact with feral cat predation should be considered, such as habitat fragmentation, clearing of trees and brush and direct human impacts.

Eradication of cats is not a practical approach for mainland areas; eradication from islands has been achieved at great cost and requires a variety of methods as well as considerable time. Most baiting methods have been ineffective due to cats failing to ingest the bait (Risbey *et al.* 1997). Several small studies using secondary poisoning of predators with agent 1080 (sodium monofluoroacetate) or brodifacoum (a second-generation anticoagulant) via poisoned prey species suggest that this may be a more effective method to kill all predators present, including feral cats, stoats and ferrets (Short *et al.* 1997; Gillies & Pierce 1999; Alterio 2002).

For example, feral cats and rats were eradicated on Fregate Island in the Seychelles using brodifacoum bait drops (Shah 2001). This was possible because there were no native mammal species and all endemic birds at risk of accidental poisoning were caught and held in captivity during the baiting, since the island was so small.

Cats on Marion Island were originally introduced in 1949 to control house mice (Bester *et al.* 2002). By the mid 1970's they were believed to be causing a decrease in bird populations, so an eradication program was devised. It included biological control with feline panleukopenia virus, hunting, trapping and poisoning during a 15-year period following four years of study and planning (Bester *et al.* 2000; Bester *et al.* 2002). This demonstrates the intense effort required to eradicate cats, even in a closed population.

Eradication of cats on Little Barrier Island, New Zealand was carried out from 1977 to 1980 using cage traps, leg-hold traps, dogs and 1080 poison (Girardet *et al.* 2001). Only leg-hold traps and poison were found to be effective; 151 cats were killed, as were some birds and rats.

Eradication of 30 cats on Gabo Island included shooting, trapping and 1080 poison bait programs (Twyford *et al.* 2000). Only the poison bait was considered to be effective, and cats were eradicated from the island after four years.

On Dassen Island, South Africa, cats were studied to determine the effects of their predation on native birds (Apps 1983). Following culling, cat numbers rapidly rebounded because some breeding cats remained, leading to re-population of the island with young cats.

Models to evaluate the efficacy of eradicating cat populations on islands using feline leukaemia virus (FeLV) and feline immunodeficiency virus (FIV) have suggested that the former could be effective if the natural immunity of the population is low (Courchamp & Sugihara 1999). However, there are many considerations when introducing a disease into an environment, such as the susceptibility of non-target species, the performance of the pathogen in the field and host susceptibility. A virus-vectored immuno-contraception approach for controlling cat populations has been modeled using parameters appropriate for islands (Courchamp & Cornell 2000). Control or eradication of the cats was deemed to be possible, if the assumptions in the model were correct regarding baiting rate, transmission rate, mortality, and determinants of population growth. Concerns about virus-vectored approaches include effects on non-target species, public acceptance of genetically-engineered organisms, spread of the vector outside the targeted location, irreversibility, genetic changes in the target species or vector, rate of response to exposure and limited knowledge of potential vector candidates (Courchamp & Cornell 2000).

5.2   Trap and Remove

In the United States, Canada, and Europe, feral cats are most often trapped and removed. What happens to them after removal varies widely and is a matter of debate. Usually, they are destroyed since they cannot be placed as companion animals. Most animal control agencies (government-run organizations) euthanize feral cats that enter their facilities. Some have mandatory holding periods while others determine that the cat is feral on arrival and euthanize it shortly thereafter. A few have programs that place cats with local feral cat organizations. Many non-profit (non-governmental) organizations do not accept feral cats unless they have a special program to deal with them.

Euthanasia may be the best option for feral cats that are injured or very ill, since long-term veterinary care is usually not possible. Intensive removal programs, with adoption of kittens and socialized adults and euthanasia of feral cats that cannot be relocated or socialized, may be an option in geographically isolated areas where predation clearly threatens native species in decline, or in areas that are unsafe for cats. This type of program must have a strong educational component and commitment to adoption in order to be accepted by residents, and must also include ongoing monitoring for the immediate removal of new cats.

In the past decade in North America, there has been an increasing tendency to recommend removal and relocation of cats to another property, often a rural home, farm or sanctuary. Sanctuaries are facilities that hold animals, often for the rest of their lives, and they may also have adoption programs. They are expensive to run well and require careful planning to provide for the needs and health care of cats throughout their lives. In the United States and other countries, there is limited oversight of the quality of care and housing provided for animals in sanctuaries and the conditions in some may be poor (see Chapters 5 and 7). While relocation to a farm setting or placement in a high quality sanctuary are attractive solutions, they are not practical as the sole solution due to the large numbers of feral cats and the limited funding available. In special circumstances, well-run sanctuaries coupled with ongoing trapping may provide a local solution. Relocation may also be one component for the control of feral cat populations in conjunction with other approaches.

5.3 Trap, Neuter and Return

Trap, neuter and return (TNR) programs in their simplest form include the humane trapping of feral cats, sterilization by a veterinarian, vaccination for rabies in countries where that is appropriate, and return to the site of trapping. Before release back into the colony, the ear of the cat should be tipped or notched to indicate that it has been sterilized (Cuffe *et al.* 1983) (Figure 3).

Figure 3. The distinctive silhouette of an ear-tipped cat is easily identified and indicates that the cat has been sterilized.

The aim of a TNR program is to create a stable population where cats can no longer reproduce; natural attrition will eventually decrease numbers or at least maintain a stable number of cats. Since cats are returned to the original habitat, a vacuum is not left to encourage cats from nearby areas to move in or remaining intact cats to repopulate. Because there is always the potential for cats to join the colony, the program must continue to trap new cats that migrate into the area. An aggressive adoption program for tame adults and kittens under about eight weeks of age will reduce the numbers of cats in the colony more quickly (Levy *et al.* 2003). Sterilization decreases roaming of male cats, improves body condition (Scott *et al.* 2002) and tends to make cats more interactive with their caretakers (Scott & Levy 2003). Thus, TNR together with adoption and monitoring programs are the most effective and humane options for the long-term control of feral cat colonies. TNR also retains the positive aspects of the presence of cats in specific locations. These include rodent control, especially in cities and around houses and barns, the opportunity to learn about cat behaviour and social interactions, the aesthetic benefits of cats in the urban environment and the relationships between the cats and their human neighbors and caretakers (Natoli 1994) (Figure 4).

Figure 4. Over time, feral cats can become more sociable with a caretaker who is familiar to them.

I am aware of three locations, two in the United States and one in England, where TNR programs and coordinated efforts to address the sources of feral cats led to the disappearance of colonies. Although about ten years was required for this to occur, this demonstrates that TNR is a humane and successful management technique for the feral cat population (Remfry 1996).

Extended programs are referred to as TTVAR-M: trap, test (the cat is blood tested for a range of diseases), vaccinate (often against a number of diseases), alter (neuter), return (to the original location) and monitor (including regular feeding by a caretaker). Cats are blood tested to see if they are infected with FeLV or FIV. This testing is controversial, as costs are high and cats positive for these viruses are usually euthanized or placed in sanctuaries. Placing virus-positive feral cats in sanctuaries is difficult, as most sanctuaries have so many healthy socialized cats needing homes that it may not be practical to spend resources on these feral cats. Testing should not be performed if no action is to be taken for virus-positive cats. There are also other reasons not to test. The prevalence of FeLV and FIV in feral cats is usually quite low, as low or lower than that found in owned cats (Lee *et al.* 2002). FeLV is spread by prolonged close contact between cats and from mother to kittens, and is not highly contagious. FIV is spread by biting during fighting, particularly among males, and its transmission is curtailed when cats are sterilized. FIV infection has a different natural history than FeLV in that infected cats can often live for a normal lifespan (see Chapter 8). Control programs usually decide to test based on the opinion of their veterinarians and on the trade off of costs and benefits. Whether cats are vaccinated for disease other than rabies will depend on the program. Most feral cats are likely to have been exposed to the common infectious diseases. Vaccination may be performed in order to protect the organization from negative comments, since their feral cats will be as well protected against viral diseases as pet cats.

5.3.1 TNR Programs Internationally

TNR appears to have originated in South Africa and Denmark well over two decades ago (Kristensen 1980; Tabor 1983). It was then imported into England and from there to the United States, Canada, Europe and many other countries (Remfry 1996). Because many programs are small and local, it is impossible to quantify the extent and success of TNR in most locations. Using networks of animal protection contacts and web sites, as well as published studies, I have collected some information to give a sense of what is happening internationally in several locations. This is not a comprehensive listing but is based on expert opinion shared with me.

5.3.1.1 TNR Programs in the United States

In the United States, TNR has become an established approach in some locations and has been on the national and regional agenda of governmental and non-profit organizations since the early 1990's. In the late 1990's, most of the large animal-protection organizations, as well as the national veterinary organization, acknowledged the usefulness of TNR, at least under certain specified circumstances. At the same time many bird, wildlife and public health organizations developed policy statements against TNR, primarily because of concerns regarding predation, rabies and lack of data on efficacy. Because laws governing cats are usually made at the local city level, general statements about the acceptance of TNR are not possible. Several large programs in the northeast and west have become increasingly high profile in animal welfare and animal protection conferences and web sites, indicating a growing awareness, if not always acceptance, of TNR as a humane method for the control of feral cats. Alley Cat Allies is a national organization dedicated to TNR for feral cats with over 90,000 donors and supporters in 2003.

An early study of a TNR program was conducted on hospital grounds in Louisiana and reported on both efficacy (control of cat numbers) and longevity of cats (Zaunbrecher & Smith 1993). Of the 41 cats present at the start of the study, 40 were returned to the site. During three years of follow-up, five cats died, five disappeared and six joined the colony. Litters of kittens were not reported during the study, and two cats became more social with the people feeding them. Beginning in 2002, the efficacy of TNR in the Unites States began to be described in the scientific literature. One was a campus TNR program in Texas where the first two years of data were presented (Hughes & Slater 2002). During that period 158 cats were trapped (Figure 5), 101 were returned and 32 kittens and tame adults were adopted. The number of kittens trapped decreased significantly between the first and second year, as did the number of complaints to the university pest control service. During the following three years, the number of trapped cats continued to decrease. Totals for the five years of the program were: 226 cats trapped, 105 returned to campus (Slater 2003). Of those returned, 15 were eventually adopted and seven were killed or died. No kittens were born on campus after the second year and fewer than 20 cats were trapped in each of the last two years of the program, with almost half being tame cats or kittens.

Another campus program in Florida documented the effect of TNR with an adoption program during an 11-year period (Levy *et al.* 2003a). A total of 155 cats were recorded during this period. After the first five years, 68 cats were present on campus and six years later 23 were present. The final disposition of all cats was: 47% were adopted (including more than 50% of cats that were initially considered feral), 15% remained on campus, 15% disappeared, 11% were euthanized, 6% died and 6% moved to nearby woods. No kittens were found after the first five years of the program.

A study of 132 colonies in Florida found that the total population of cats decreased from 920 to 678 after TNR (Centonze & Levy 2002). Median colony size was initially four cats (range one to 89), and was reduced to three (range zero to 42) following TNR. The greatest source of new cats was births, and adoptions led to the greatest decrease in numbers.

An animal control agency serving a large county in Florida initiated a TNR program in collaboration with a local feral cat organization in 1995 (Hughes *et al.* 2002). Six years of data before and after the implementation of TNR for feral cats demonstrated that there was no increase in complaints or impoundments by the animal control agency.

Figure 5. Humane box-traps are commonly used to capture feral cats. Cats will usually become very agitated after the trap closes, so the door should be securely latched and the trap covered immediately to reduce the cat's stress level

During the study period, the human population increased by a third, which should have led to one third more cats, cat-related complaints, impounds and euthanasias. In fact, euthanasia rates and complaints decreased during the last five years. Numbers of sterilizations increased dramatically in the six years after TNR and low cost sterilization programs were instituted for feral and owned cats. The relationship between the agency and the public improved, as did the morale of the animal control officers. In addition, TNR provided concerned citizens with the option to take action and make a difference to the numbers and the well-being of feral cats in their neighborhoods.

5.3.1.2 TNR Programs in Other Countries

Some of the earliest published studies come from England, and focus on the longevity and behavioural impact of TNR programs. The behaviour and stability of the groups were studied and found to be "satisfactory on both counts" (Neville & Remfry 1984). Seventeen other neutering programs were followed for five years (Remfry 1996): a total of 253 cats were trapped, 201 were returned to their original site and 141 were still present five years later.

In Canada, several organizations dealing with feral cats exist and some research interest in the area has developed. As of early 2003 there was no national organization, but the no-kill movement (which embraces the idea that euthanasia of healthy animals is not a viable solution to overpopulation) is picking up momentum. The first national conference on the subject took place in the early summer of 2003 and attracted participants from Canada and the United States. Part of their activity was the formation of a national organization of groups and individuals working toward a no-kill policy ("Let-Live Canada").

The summary of the situation in Israel is based on two reports from the Cat Welfare Society of Israel (personal communication Rivi Mayer, May 3, 2003; personal communication, Adi Nevo, May 1, 2003). Although the feral cat population is much in evidence, feral cats are not a common concern of the public or the government. In general, cats as pets and companions are not highly valued or commonly kept (personal communication, Rama Santschi, DVM, July 17, 2003). Cats primarily come into the public and government awareness as nuisances or concern about rabies. Feral cats have been rounded up and destroyed for 50 years without making any difference to numbers. However, the Cat Welfare Society of Israel has been active since 1990, and members have made strong efforts to network and learn from existing programs about controlling feral cat populations and implementing spay/neuter programs. Recently, the Society has sterilized about 4,500 cats a year and provides a trapping and transportation service. Several cities have begun TNR programs but most have not persisted with them due to a combination of limited funding, lack of commitment by the government and shortage of structural support. Added to the complexity of the situation is the fact that, on the one hand, the Ministry of Environment administers animal rights and protection and also supports TNR, both philosophically and financially. On the other hand, the Department of Veterinary Services is part of the Ministry of Agriculture and tends to promote lethal methods to control cat populations. City-employed veterinarians are in charge of municipal animal activities and tend not to understand or become involved in TNR or subsidized sterilization programs. Nationally, the Supreme Court determined in 1997 that the mass killing of dogs and cats was not permitted, and that each animal-related complaint must be evaluated. The Court also declared that non-lethal solutions, including TNR, should be sought, and refined the rules regarding the control of rabies. Unfortunately, this has not stopped some private trappers from continuing to trap and kill cats under regulations from the Ministry of Agriculture. Although the situation has improved with more sterilization programs, less killing of cats and increasing awareness of TNR as a solution to the problem, funding and veterinary support continue to be limiting factors. A recent article describes how the existing literature on free-roaming cats can be applied to the situation in Israel (Gunther & Terkel 2002). The conclusions of the authors were to promote trap, neuter, identify and return programs in conjunction with community level solutions like keeping garbage cans securely covered, education and dealing with specific problems. They recommended trap and euthanasia only for cats in very poor condition.

In Germany, a recent dissertation on feral cat populations in a 45 ha study area in Berlin was completed by Beate Kalz (Edoc.hu-berlin.de/abstract.php3/dissertationen/kalz-beate-2001-02-28).In her opinion, feral cats are not a highly visible group and are generally well tolerated by the public. German animal welfare organizations usually promote TNR as a control method, with a strong emphasis on the sterilization of cats.

The Dutch Society for the Protection of Animals has been in operation for 130 years. The 110 shelters in Holland are all associated with the Society. Between 1992 and 1996, the numbers of stray cats in the shelters increased by nearly one third, to 31,100. In 1997, 15,000 owned cats, 3,000 feral cats and all cats in shelters were sterilized (a total of 48,800 cats). In the late 1990's a national sterilization campaign was developed and implemented. This information would suggest that TNR is practised fairly widely and successfully in Holland, and that its administration has benefited from the long history of animal welfare activities in that country.

An estimated 80,000 stray cats live in Singapore. In May 2003, Singapore's Agri-food and Veterinary Authority (AVA) used the Sudden Acute Respiratory Syndrome (SARS) outbreak as a reason for the intensified culling of stray cats, especially in areas with nuisance problems (The Straits Times, Singapore, May 23, 2003). The Society for the Prevention of Cruelty to Animals and other welfare groups countered this with a call to end the culling of cats. The following day, the AVA reversed its position and denied a link between the culling and SARS, and indicated that it was for other public health reasons. In 1998, AVA's Stray Cat Rehabilitation Scheme had sterilized about 5,000 cats through their own Cat Welfare Society, but with the initiation of culling this sterilization program was put on hold. Animal welfare organizations continue to seek to relocate cats to sanctuaries and end the culling of cats. While TNR had been implemented in Singapore, it has not been adopted at a level to decrease cat-related complaints significantly. Furthermore, Singapore's example illustrates that even when there is a government program for TNR, its position may revert to old methods of removal and euthanasia in large numbers without good reason.

No government office oversees animal welfare at the national level in Japan (Oliver 2002). Free-roaming dogs and cats are collected and disposed of by the Department of Health & Hygiene. Cats have only recently begun to be regarded as companion animals rather than as working hunters. Following the control of free-roaming dogs Japan, like many other countries, now has more obvious colonies of feral cats but TNR is rarely practised. One colony of feral cats has been extensively studied, for example see Izawa (1983), Yamane *et al.* (1997) and Ishida *et al.* (2001).

In 2000, a staff member of the Hong Kong SPCA introduced TNR to Hong Kong (Garrett 2003). In just over three years, 2,200 street cats have been sterilized and cared for (about 100 cats a month in 2003), with 15 registered cat carers and 60 part-time carers. The SPCA provides free spay and neuter, vaccination and flea control services, and also has a mobile clinic that provides similar services to villages and islands. In August 2001, it declared its intentions to make the city of Hong-Kong adopt a "no-kill" policy.

These examples demonstrate the range of views about killing and caring for cats, as well as differing perceptions of what the feral cat problem, and its control, entails. They also support the slowly evolving view that feral cats are worthy of our concern and compassion.

6. CONTROLLING THE SOURCES OF FERAL CATS

In the previous section, I briefly reviewed the methods of controlling free-roaming cat populations. Particularly because of public health and wildlife concerns, the choice of control method can be controversial. When one has considered the financial costs, the welfare of the cats, the need for solutions tailor-made for each location, and a shortage of data on the efficacy of different methods, the choice may not be obvious. Nevertheless, the sources of these cat populations also need to be addressed.

One often hears the phrase "responsible pet ownership". It implies that a certain level of care is due to companion animals. Responsible pet ownership includes the provision of suitable food and shelter, health care and social interaction, and, I believe, the permanent identification of the animal (a tattoo or microchip), the provision of a safe environment and a life-long commitment to the animal's care. The community should view abandonment not only as a failure of individual responsibility but also as an antisocial and immoral act.

There are a number of approaches to reducing the number of cats entering the feral cat population. Firstly, stray cats need to be reunited with their families; in the United States, only 2 to 3% of all cats entering shelters are returned to their owners (Zawistowski *et al.* 1998; Wenstrup & Dowidchuck 1999). The reunification rate is improved substantially in locations where major microchipping and identification programs of cats have been implemented (Slater 2002). Secondly, cats allowed outside should be sterilized and thirdly, owners should seek help for behavioural, medical or pet selection problems. Many owners do not keep their cats long-term because of a lack of knowledge about normal cat behaviour and social needs (New, Jr. et al. 2000). They may relinquish a cat to a shelter after living with its behaviour problem for years, rather than seeking help early on when the situation could be improved (DiGiacomo et al. 1998). Subsidized sterilization should be available for those who cannot afford full-cost services, and owners should be helped to find homes for cats they cannot keep. Leadership at both national and local levels is needed (Christiansen 1998). Components of community-based programs should include: 1) public education from pre-school to adult; 2) improving the quality of animal control; 3) developing expertise in urban animal management; and 4) understanding companion animal population dynamics (Murray 1992).

There may also be a role for legislation to prevent owned cats from becoming part of the feral cat problem. However, some forms of companion animal legislation may have drawbacks. A law against abandoning cats seems logical, but would be very difficult to enforce. Such a law could be construed to include TNR programs, that is, caretakers returning sterilized cats to colonies could be accused of abandoning them. While legislation, if thoughtfully written and enforceable, is likely to be beneficial, it should be adapted to enable TNR programs to continue.

7. FERAL CAT ISSUES

When considering the management of feral cat populations, the effects of predation of wildlife by feral (and non-feral) cats, public health issues (such as zoonotic disease) and the welfare of the cats themselves are major concerns that should be addressed.

7.1 Effects of Predation on Wildlife

The effect of predation of wildlife (mammals and birds) is probably the most controversial issue regarding feral cats. Unfortunately, the discussion about cats and wildlife is often polarized and couched as pro-cat versus anti-cat, or as pro-cat versus pro-wildlife. This division is inaccurate, misleading and counterproductive; in fact, there are many points in common and much overlap between the "cat" groups and the "wildlife" groups. For example, suggestions for reducing predation of wildlife by cats that is often espoused by both cat and wildlife organizations includes keeping cats indoors or confined, sterilizing cats, improving the environment for birds and bats with nesting boxes and carefully considering bird feeder placement (Gray 1999).

There are several themes that arise in discussions of feral cats and wildlife. The first is based on a philosophical belief that since cats are a domestic species, they should not be allowed to hunt wildlife but should be confined indoors, to an enclosure or yard or on a leash (arguably for the cats' welfare as well as for that of wildlife).The second theme is that cats are an introduced, non-native species and therefore should be removed from the environment. There are several assumptions underlying this argument: firstly, introduced or non-native species are harmful and native species should be protected from them. However, cattle and sheep are routinely protected from coyotes, foxes and wolves, despite the latter being native species that are killed because they may prey on domestic species (Cohen 1992). In some locations, native mountain lions, northern harriers and kestrels have been killed to prevent them from preying on rare species (Cohen 1992). The second assumption is that if we remove cats from the environment, the ecosystems will return to "normal" or to the pre-cat situation. However, ecosystems are complex and have often been heavily influenced by the effects of human habitation including construction, changes in fire control and water movement, pollution and the introduction of livestock. There are often other introduced plant and animal species (starlings or rats) that affect the balance of the ecosystem. For example, removing cats in certain locations may cause serious problems from the resulting increases in rodent populations. The third theme is the actual impact of cats on wildlife, largely through predation but also through competition or disease. While competition is commonly cited as a concern, little evidence is available to support this claim (George 1974). Predation is generally considered to be the most serious problem, especially predation of birds. Again, the interaction between cats and wildlife varies widely from location to location, and is heavily influenced by other environmental factors such as variety of prey species, the reliance of cats on garbage or being fed, other pressures on local species, climate and the biology of threatened species.

Predation is often studied by examining the diet of cats in different locations; an excellent review of such studies can be found in Fitzgerald & Turner (2000). Methods of quantifying the diet of feral or free-roaming cats include examining intestinal samples from cats that are killed, scat (faeces) analysis, recording prey brought home by owned cats and examination of dead or partially eaten prey found in the environment. The results of diet studies do not provide evidence of the impact on a species unless prey species abundance is also monitored, as well as the species' reproductive capacity and other sources of predation and mortality (Churcher & Lawton

1987; Martin *et al.* 1996; Risbey *et al.* 1999; Edwards *et al.* 2000). While predation patterns in a given location are unique, there are some generalizations that can be made. On continents mammals are the main prey eaten by cats, with birds forming about 20% of the diet (Fitzgerald & Turner 2000). The amount of household food available to cats will depend on the density of the human population. Australian cats living near refuse dumps and towns were found to have food scraps as a high proportion of their diet, while the diet of those living distant from human habitation contained few food scraps (Risbey *et al.* 1999).

Relatively few species of mammal commonly form most of the diet. Birds are a less frequent component of the diet, but usually many more species are eaten. The number and species of reptiles as food items will vary widely among locations.

7.1.1 Pro-cat versus Pro-wildlife

Some believe that to allow owned cats loose to hunt, or to maintain free-roaming cat populations in the natural environment, places more value on the life and needs of the cat than on the life of the prey the cat kills. This argument is a personal ethical belief about the relative importance of different non-human animals, rather than concern over reductions in prey species. Cats sometimes precipitate this belief by presenting their owners with prey (Dunn & Tessaglia 1994). Sweeping generalizations are often made about cat predation and are not always based on the offered evidence (Gray 1999). Additionally, data are often extrapolated inappropriately (Dunn & Tessaglia 1994). Studies that count the number of prey returned to owners are subject to many kinds of biases. Owners of cats that are better hunters are more likely to volunteer for prey studies (Fitzgerald & Turner 2000). Relatively few cats bring in very large numbers of prey, skewing the results and artificially inflating the mean number of prey; using the median would be a more suitable measure.

One example is a commonly discussed one-year study of prey brought home by 70 owned cats in an English village (Churcher & Lawton 1987). There was an average of 14 prey per cat (range zero to 95), the median was not presented but, based on a graph, seemed to be eight. Mammals comprised most of the prey (mainly wood mice, voles and shrews) and birds about 35% (mainly the house sparrow). The age of the cat (older cats brought home less prey) and their location in the village influenced prey numbers. Cats were estimated to account for at least 30% of sparrow deaths in the village and were considered to be the major predator of house sparrows. However, there was an unusually high density of sparrows in the village and other predators were not assessed. In addition, there was no indication that this level of predation had caused the sparrow population to decline.

In a questionnaire study involving 1,300 rural residents in Wisconsin (Coleman & Temple 1989), a fifth of the 800 respondents did not have cats. The remaining owned between one and 60, with an average of five cats per farm or rural residence. They reported 279 prey captures on the 20 to 30 farms and residences in the study area, with mammals making up 68% and birds 23% of the prey. These figures were used as the basis for an article with the headline "cats kill millions of small mammals and birds every year" (Harrison 1992).

Even some who value wildlife over cats will acknowledge that there are certain wildlife species that are pests which could be controlled by predation, and that using cats to control rodent populations around barns or stables is generally acceptable. Endangered species are rarely encountered in urban environments, and there are often large numbers of introduced prey species. In these settings, feral cats may be useful in controlling rodents and introduced species, and are likely to have little impact on endangered or declining species. The large population of some birds and pests in urban environments has been attributed to a variety of factors including a reduced number of predators, favourable microclimates and/or food availability (Sorace 2002). Studies of three Italian parks found high prey (including pest species such as pigeons, starlings, mice and rats) and high predator (birds of prey, crows, cats, dogs, rats and foxes) densities compared to the nearby countryside (Sorace 2002). The numbers of nest predators such as blue jays, raccoons and opossums, and of bird species that lay their eggs in nests of other species, often grow in urban environments due to a proliferation of food supplies (Terborgh 1992). In addition, current "garden" or suburban birds may be under less predation pressure from cats than they would be from the range of native predators that no longer co-exist close to human habitation (Mead 1982).

7.1.2 Cats as an Introduced Species

Invasive or introduced species are a growing concern in many countries, including the United States (Dinsmore & Bernstein 2001) and Australia (Burbidge & Manly 2002), and cats are considered to be an introduced species. Introduced carnivores can affect the local species by competition, predation, interbreeding or disease (Dickman 1996a; Courchamp *et al.* 1999; Macdonald & Michael 2001). While these processes affect individuals, effects at the population or community level may or may not occur (Dickman 1996a). Usually cats are only one of many introduced species including rats and the dogs, mongooses and weasels that were released to control them (Jackson 1978). In addition, the livestock species that were brought in such as pigs, sheep, cattle and goats may also cause serious changes in the environment, especially in the large numbers associated with industrial farming (Jackson 1978).

Being an introduced species, cats are often targeted for control measures even when there is little evidence to support this. For example, on Socorro Island, Mexico, the Socorro Mocking bird had declined in numbers (Martinez-Gomez et al. 2001). Habitat destruction was considered to be the primary cause, since Northern Mockingbirds and cats arrived after much of the decline had occurred. Nevertheless, cat control was still a major focus of the authors. Reports blamed cats for the disappearance of three petrel species on Little Barrier Island, New Zealand (Veitch 2001), yet no evidence exists that these species were ever present (Girardet et al. 2001). Only in the past few years have predators other than cats, such as ferrets and stoats, been considered in studies of predation in New Zealand (Moller & Alterio 1999; Gillies et al. 2000; Norbury 2000).

7.1.3 Extinction of Native Species

Habitat destruction by humans generally takes three forms: over-exploitation of resources, pollution and introduction of exotic species (Macdonald & Michael 2001), and is generally considered to be the most important cause of species extinctions (Lawren 1992; Terborgh 1992; Hall et al. 2000; Dinsmore & Bernstein 2001; Macdonald & Michael 2001). Water quality deterioration, drainage of wetlands, agricultural use of prairies, fertilizers, pesticides and herbicides are all responsible for changes in the environment of a variety of bird habitats, which lead to declines in populations (Terborgh 1992; Robinson 1998). It is crucial to view cat predation within the context of habitat destruction, since cats have not been shown to be the primary cause of the loss of native species on mainland continents (Mead 1982; Mitchell & Beck 1992). Unfortunately, evidence regarding extinctions is often anecdotal, circumstantial or historical (Dickman 1996a; Macdonald & Michael 2001; Read & Bowen 2001).

Islands have less species diversity, a scarcity of predators and a higher concentration of individuals relative to similar mainland environments (Sorace 2002). Islands with introduced cats differ enormously in climate, size and native species, but generally have relatively few native mammals (Fitzgerald & Turner 2000). The same set of introduced species is common: house mice, rats and European rabbits. Where rabbits are present, they tend to be the main prey of cats. Predation on rats and mice varies between locations. Cats survive on islands without mammals by eating seabirds on small islands and land birds on larger islands.

Australia is arguably the best studied and most high profile country when it comes to feral cats and predation, and is considered to be an example of the serious threat that feral cats pose to wildlife. However, as of 1995, there were "no critical studies of the impact of feral cats on native fauna in Australia" (Dickman 1996b). What has been documented is the association between rainfall, species' habitat and dietary preferences, and the decline and extinction of species (Burbidge & McKenzie 1989). European settlement led to a reduction in vegetative cover, increased human settlements and introduced species, including livestock, and changes in control of fires in the environment. Exotic predators likely exacerbated the situation, depending on the protective habitat of the prey species. Feral cats are not recorded to have had a significant impact on any species of reptiles, amphibians, fish or invertebrates (Dickman 1996b); however, they may have localized effects on populations of native vertebrates. Despite much publicity, the role of feral cats in the decline and extinction of Australian mammalian species remains unclear (Burbidge & Manly 2002)

Finally, in addition to direct predation, there have been concerns about diseases that could be spread from cats to wildlife. The 'Alala bird in Hawaii became endangered possibly due to disease, loss of genetic diversity, introduced predators or habitat loss (Work *et al.* 2000). Reintroduction programs were limited by the presence of the microorganism *Toxoplasma gondii* (for whom the cat is the main host) in four of 27 captive-reared birds due for re-introduction. It is unclear if these particular birds or 'Alala birds in general are especially susceptible to toxoplasmosis, perhaps due to a genetic predisposition. Toxoplasmosis was suggested to be a contributing factor to local decreases in eastern barred bandicoots in Australia (Dickman 1996b).

7.1.4 Complexity of Ecosystems

In order to understand their role in predation, it is crucial to recognize that cats are one of a large group of predators, both native and introduced, (Fitzgerald & Turner 2000). Other introduced species, such as rats and mice, can have substantial impacts on amphibians, mammals and birds (Courchamp *et al.* 1999). Many factors affect the impact of cats on prey species, such as the density of cats, the density and distribution of prey, the fecundity of native species, the habitats and habits of native species, and the presence of other predators. Assumed relationships may not be correct when studied over long periods of time (Fitzgerald & Gibb 2001).

Because the relationship between different predators and a variety of prey species is complex, removal of cats may have much more widespread effects than are immediately obvious. This is illustrated by a mathematical model including birds, rats and cats, which showed that removing all cats led to a surge in rat numbers, resulting in the extinction of the bird species (prey) (Courchamp et al. 1999). Another model examined the relationship between birds (prey), rabbits (an introduced prey species) and cats (the predator) in an island setting (Courchamp et al. 2000). Based on field observations, rabbits provide food for other predators and, in times of plenty, are the primary diet of cats. This allows for a larger population of cats than could ordinarily be sustained if rabbits were scarce or not present. When rabbit populations are reduced, cats are able to switch to other prey species (such as birds). Similarly, the widespread availability of cat food could lead to larger populations of cats than would otherwise be possible if only local prey were available.

Despite the eradication of cats on Marian Island, lesser sheathbill populations remained less abundant and had different habits than birds on neighboring Prince Edward Island (Huyser et al. 2000). These differences were believed to be due to a decrease in the birds' macro-invertebrate prey (especially weevils and flightless moths), which may have been due to increases in house mice as a result of the cat eradication, decreases in burrowing petrels (which promote invertebrate species), and climate warming, which also increases mouse populations. This example illustrates that the removal of cats may not result in the recovery of a threatened species.

7.2 Public Health and Zoonotic Disease

Most of the agencies charged with public health issues are concerned with the possibility of disease rather than with the actual probability, particularly in regard to cats. This is partly due to the lack of data regarding frequencies of zoonotic diseases and the risk of transmission.

Rabies in cats is often the chief concern of public health authorities, especially in countries where the disease is common. There are many other zoonotic diseases where cats are implicated (Tan 1997; Patronek 1998; Olsen 1999). Some of them are region-specific, such as plague in the Western United States (Orloski & Lathrop 2003) and others may affect cats as well as many other mammalian species (Riordan & Tarlow 1996). While any free-roaming or owned cat may carry or transmit a variety of diseases to humans, the frequency of these diseases and their severity will fluctuate widely depending on the geographic location, climate and the health status of the human population.

Proper handling of feral cats, using traps and other equipment, will reduce the likelihood of bites and scratches, thereby reducing the risk of disease transmission (Slater 2000). Although cat bites in the United States are less common than dog bites, they are more likely to become seriously infected because of the micro-organisms present in cat saliva (Tan 1997).

In southern Africa, parts of the Caribbean, North America and Europe, wild carnivores are the primary vector for rabies, while in Asia, parts of Latin America and most of Africa, dogs continue to be the major source (WHO 2002a). The United States is the only country where cats were the most commonly diagnosed domestic species in recent years, yet cases of laboratory-confirmed rabies in skunks, raccoons and bats in the United States far exceeded the numbers of all domestic animal species combined (WHO 2002b). Historically, measures such as quarantine (restricting animal movements), removing free-roaming animals and vaccinating susceptible animals have been used to control rabies (Beran & Frith 1988). Originally, susceptible animals included only domestic species but in the 1980's wildlife species also began to be vaccinated, using oral bait systems.

Relatively little research has been done on cat populations and the control of rabies, although dogs have been studied in a number of countries and some solutions have been devised (WHO Expert Committee 1988; Meslin *et al.* 1994; WHO Expert Committee 1994). Cat population dynamics are likely to parallel those of dogs in many locations, so similar solutions will be effective.

Feral cats should be vaccinated for rabies in locations where rabies occurs, and vaccination of colonies will result in a herd immunity effect. Herd immunity is the point at which the proportion of immune individuals in the group is so high that the disease agent cannot enter and spread (Hugh-Jones *et al.* 1995). A level of 80% immunity among dog populations is sufficient to break the transmission cycle of rabies (WHO 2002a). In 1999, health officials in Ontario, Canada, incorporated the vaccination of free-roaming cats into their emergency response to outbreaks of rabies in raccoons (Rosatte *et al.* 2001). All cats within ten km of the initial raccoon rabies case were trapped and vaccinated. During this outbreak about 800 cats were vaccinated instead of killed, and provided a partial barrier to disease spread.

Toxoplasmosis is another widely-occurring disease in cats that is transmissible to humans. The acute infection is generally self-limiting in immuno-competent humans, but may cause serious disease in immuno-compromised humans (AIDS patients in particular) or to the foetus during pregnancy (Schantz 1991; Olsen 1999). An additional concern is environmental, with microorganisms contaminating water or feed. The prevalence of Toxoplasma infection in feral cats appears to be similar to that in owned cats (DeFeo *et al.* 2002).

Cat scratch disease, caused by *Bartonella henselae*, has a wide range of prevalence in owned and feral cats, from zero in Norway to over 50% in the

United States and Philippines (Barnes et al. 2000; Bergh et al. 2002). There is also variable prevalence in feral cats in the United Kingdom, from 0 to 100% depending on location (Barnes et al. 2000). This disease is primarily a problem in immuno-compromised humans (Hugh-Jones et al. 1995), and requires a scratch or bite for transmission. Zoonotic diseases are also described in Chapter 3.

7.3 Feral Cat Welfare

Only in recent decades has the welfare of feral cats themselves emerged as an important issue. In a few countries it is the primary concern, while in others it remains the focus of small groups or individuals concerned with animal welfare.

Concern for the well-being of feral cats should consider not only their health but also their need for some interaction with humans. Cats in managed colonies appear to be in good health and are able to obtain whatever level of interaction they need with their caretaker. Caretakers themselves often have a strong bond with their feral cats (Haspel & Calhoon 1993; Natoli et al. 1999) (Figure 6). A study in Hawaii of 75 colony caretakers found that most were female, middle-aged, married and well-educated, owned pets, and were employed full-time (Zasloff & Hart 1998). The caretakers spent considerable time and money caring for these colonies because of their love of cats and the opportunity to nurture them. They also experienced enhanced feelings of self-esteem. A second study in Florida of 101 caretakers of 920 cats in 132 colonies found that 84% were female (Centonze & Levy 2002). The median age was 45 years (range 19 to 74 years) and 88% owned pets (two-thirds of them owned cats). More than half the caretakers were married. The most common reason reported for caring for the cats was sympathy or ethical concern followed by loving animals or cats.

FeLV and FIV viruses are the infectious diseases most frequently studied in cat populations, both because of their impact on cats' health and the risk of transmission to other felines. A total of 516 stray cats (467 were classified as tame and 49 as feral or semiferal) entering an animal shelter and veterinary hospital in Birmingham, England, between August and December 1997, was tested for FeLV and FIV (Muirden 2002). In all cats, the prevalence of FeLV antigen was 3.5% and of antibodies to FIV was 10.4%. The prevalence of FeLV in semiferal or feral cats (2%) was similar to that in tame cats (3.6%), while the prevalence of FIV was 2.5 times higher (20.4 versus 9.4%). There were also higher rates of FIV antibody-positive status in males, cats over two years of age and cats with non-traumatic health problems. Multivariate analysis indicated that sex, age and non-traumatic illness were independently associated with FIV antibody-positive status but feral status was not.

Figure 6. Caretakers may spend hours traveling to their colonies to feed, nurture and interact with the cats.

Another study of FeLV and FIV in veterinary practices in Istanbul, Turkey, included indoor cats, cats allowed outside and feral cats (Yilmaz *et al.* 2000). The latter two groups were combined for analysis, which makes reaching conclusions about the feral cats difficult. Prevalence of FIV in both groups was 22% (9/40 indoor and 14/63 outdoor cats) and of FeLV was 5% in indoor and 6% in outdoor cats. FIV was more common in male cats; the high prevalence may be related to the fact that most cats were not neutered. The indoor cats may have been previously outdoor cats or from the same household as some of the outdoor cats, which could bias the infectious disease frequency, but no data were given. These studies demonstrate the variability of disease prevalence in different populations of owned and feral cats, and the difficulty in making comparisons between studies that define cat populations differently.

Among 226 cats trapped during five years of a Texas university campus program, 5% were positive for FeLV and 6% for FIV (Slater 2003). None of the cats trapped were euthanized for other serious health problems. In the Florida university campus program, 11% of cats were euthanized for serious illness (Levy *et al.* 2003a). Of these, 7% were positive for FeLV or FIV.

Operation Catnip, a high-volume spay/neuter program for feral cats (Figure 7) in Florida and North Carolina found that 4% of 733 cats were positive for FeLV or for FIV (Lee *et al.* 2002). FIV was more common in males. Among a larger sample (5,766) of cats from Operation Catnip, nine cats were euthanized for serious health problems (other than FIV and FeLV) and 17 died from apparent anesthetic complications (nine had physical abnormalities that may have contributed to their deaths), giving a mortality rate of 0.35% (Williams *et al.* 2002). A program on Prince Edward Island, Canada, trapped and tested 185 cats and kittens during a 14-week period (Gibson *et al.* 2002). Prevalence of FeLV was 5%, of FIV was 6% and three cats were positive for both viruses; as in previous studies, FIV was more common in males. These diseases tended to occur within specific colonies, with other colonies being clear of infection. These studies demonstrate that colonies undergoing TNR tend to have few health problems and a low prevalence of FIV and FeLV, suggesting that feral cats in managed colonies, at least, pose limited health risks to other cats.

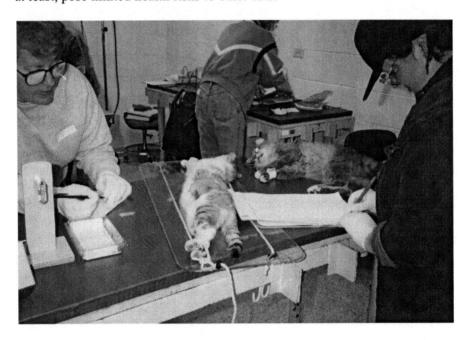

Figure 7. This female cat is being prepared for surgery at a high-volume spay and neuter clinic. In this type of clinic, over 100 cats can be sterilized in one day.

Other diseases are occasionally studied in feral cats. For example, 50 rural feral cats from a shelter near Zagreb, Croatia were examined for lungworms (*Aelurostrongylus abstrusus*) at necropsy (Grabarevic *et al.* 1999). The prevalence was 22%, much higher than the prevalence in cats seen at the veterinary college (3.9%). These populations probably differed greatly in the level of care and nutrition they received.

Weight and body condition are good clues to general health in cats. A study of body condition in 105 adult feral cats found they were lean (4 on a scale from 1 to 9) but not emaciated at the time of surgery for neutering (Scott *et al.* 2002). One year later, 14 cats were reevaluated and all of them had a substantial increase in falciform fatpad area and depth and body weight, and an increase of one level in the body condition score. Caretakers judged that their cats were friendlier, less aggressive, less inclined to roam and had improved health and coat condition.

Critics of managing colonies by TTVARM argue that feral cats live to less than five years of age and die from car accidents, disease, poisoning, abuse and attacks from other animals (Clarke & Pacin 2002). Yet the alternative for these cats is euthanasia, and the evidence presented here suggests that feral cats in managed colonies can be kept in reasonably good health and enjoy a good quality of life.

8. CONCLUSIONS

Over the past few decades and in many parts of the world, the welfare of feral cats has become a matter of great concern. This is largely due to the development of sensitivity toward animal welfare and a shift in how animals, particularly cats, are perceived. All those concerned have a common goal: fewer feral and free-roaming cats. There is often intense conflict, however, over what to do with these cats and who is responsible for them. Increasingly, there is resistance to killing cats simply because they are a nuisance, prey on wildlife or may be a threat to public health. Organizations and governments need to find non-lethal, effective, and humane methods to control feral cat populations, and comprehensive and creative community-wide programs need to address the sources of feral cats. As cats become more popular as pets and society continues to evaluate the role and care of non-human animals, the welfare of feral cats will become an increasingly central issue for individuals, societies, organizations and governments.

9. REFERENCES

Alterio, N. (2002) Controlling small animal predators using sodium monofluoroacetate (1080) in bait stations along forestry roads in New Zealand beech forest. *New Zealand J. Ecology* **24**, 3-9.

Apps, P.J. (1983) Aspects of the ecology of feral cats on Dassen Island, South Africa. *South African J. Zoology* **18**, 393-399.

Barnes, A., Bell, S.C., Isherwood, D.R., Bennett, M. and Carter, S.D. (2000) Evidence of Bartonella henselae infection in cats and dogs in the United Kingdom. *Veterinary Record* **147**, 673-677.

Beran, G.W. and Frith, M. (1988) Domestic animal rabies control: an overview. *Reviews of Infectious Diseases* **10**, S672-S677.

Bergh, K., Bevanger, L., Hanssen, I. and Loseth, K. (2002) Low prevalence of Bartonella henselae infections in Norwegian domestic and feral cats. *Acta Pathologica, Microbiologica et Immunologica Scandinavica* **110**, 309-314.

Bester, M.N., Bloomer, J.P., Bartlett, P.A., Muller, D.D., van Rooyen, M. and Buchner, H. (2000) Final eradication of feral cats from sub-Antarctic Marion Island, southern Indian Ocean. *South African J. Wildlife Research* **30**, 53-57.

Bester, M.N., Bloomer, J.P., van Aarde, R.J., Erasmus, B.H., van Rensburg, P.J.J., Skinner, J.D., Howell, P.G. and Naude, T.W. (2002) A review of the successful eradication of feral cats from sub-Antarctic Marion Island, Southern Indian Ocean. *South African J. Wildlife Research* **32**, 65-73.

Burbidge, A.A. and Manly, B.F.J. (2002) Mammal extinction on Australian islands: causes and conservation implications. *J. Biogeography* **29**, 465-473.

Burbidge, A.A. and McKenzie, N.L. (1989) Patterns in the modern decline of western Australia's vertebrate fauna: causes and conservation implications. *Biological Conservation* **50**, 143-198.

Calhoon, R.E. and Haspel, C. (1989) Urban cat populations compared by season, subhabitat and supplemental feeding. *J. Animal Ecology* **58**, 321-328.

Centonze, L.A. and Levy, J.K. (2002) Characteristics of free-roaming cats and their caretakers. *J. American Veterinary Medical Association* **220**, 1627-1633.

Christiansen, B. (1998) *Save our strays: how we can end pet overpopulation and stop killing healthy cats & dogs.* Canine Learning Center, Napa, CA.

Churcher, P.B. and Lawton, J.H. (1987) Predation by domestic cats in an English village. *J. Zoology* **212**, 439-455.

Clarke, A.L. and Pacin, T. (2002) Domestic cat "colonies" in natural areas: a growing exotic species threat. *Natural Areas Journal* **22**, 154-159.

Cohen, A. (1992) Weeding the garden. *The Atlantic Monthly* **November**, 76-86.

Coleman, J.S. and Temple, S.A. (1989) Effects of free-ranging cats on wildlife: A progress report. *Proceeding of the Eastern Wildlife Damage Control Conference* **4**, 9-12.

Courchamp, F. and Cornell, S.J. (2000) Virus-vectored immunocontraception to control feral cats on islands: A mathematical model. *J. Applied Ecology* **37**, 903-913.

Courchamp, F., Langlais, M. and Sugihara, G. (1999) Cats protecting birds: modeling the mesopredator release effect. *J. Animal Ecology* **68**, 282-292.

Courchamp, F., Langlais, M. and Sugihara, G. (2000) Rabbits killing birds: modeling the hyperpredation process. *J. Animal Ecology* **69**, 154-164.

Courchamp, F. and Sugihara, G. (1999) Modeling the biological control of an alien predator to protect island species from extinction. *Ecological Applications* **9**, 112-123.

Cuffe, D.J.C., Eachus, J.E., Jackson, O.F., Neville, P.F. and Remfry, J. (1983) Ear-tipping for identification of neutered feral cats. *Veterinary Record* **112**, 129.

DeFeo, M.L., Dubey, J.P., Mather, T.N. and Rhodes, R.C. (2002) Epidemiologic investigation of seroprevalence of antibodies to Toxoplasma gondii in cats and rodents. *American J. Veterinary Research* **63**, 1714-1717.

Dickman, C.R. (1996a) Impact of exotic generalist predators on the native fauna of Australia. *Wildlife Biology* **2**, 185-195.

Dickman, C.R. (1996b) *Overview of the impacts of feral cats on Australian native fauna.* Sydney: Australian Nature Conservation Agency, pp. 1-85.

DiGiacomo, N., Arluke, A. and Patronek, G. (1998) Surrendering pets to shelters: the relinquisher's perspective. *Anthrozoös* **11**, 41-51.
Dinsmore, J.J. and Bernstein, N.P. (2001) Invasive species in Iowa: An introduction. *J. Iowa Academy of Science* **108**, 105-106.
Dowding, J.E., Murphy, E.C. and Veitch, C.R. (1999) Brodifacoum residues in target and non-target species following an aerial poisoning operation on Motuihe Island, Hauraki Gulf, New Zealand. *New Zealand J. Ecology* **23**, 207-214.
Dunn, E.H. (1993) Bird mortality from striking residential windows in winter. *J. Field Ornithology* **64**, 302-309.
Dunn, E.H. and Tessaglia, D.L. (1994) Predation of birds at feeders in winter. *J. Field Ornithology* **65**, 8-16.
Edwards, G.P., de Preu, N., Shakeshaft, B.J. and Crealy, I.V. (2000) An evaluation of two methods of assessing feral cat and dingo abundance in central Australia. *Wildlife Research* **27**, 143-149.
Fitzgerald, B.M. and Gibb, J.A. (2001) Introduced mammals in a New Zealand forest: long-term research in the Orongorongo Valley. *Biological Conservation* **99**, 97-108.
Fitzgerald, B.M. and Karl, B.J. (1986) Home range of feral house cats (Felis catus L.) in forest of the Orongorongo Valley, Wellington, New Zealand. *New Zealand J. Ecology* **9**, 71-82.
Fitzgerald, B.M. and Turner, D.C. (2000) Hunting behaviour of domestic cats and their impact on prey populations. In Turner, D.C. and Bateson, P. (eds.). *The Domestic Cat: the biology of its behaviour*, 2nd edn., Cambridge University Press, Cambridge, pp. 151-175.
Garrett, V. (2003) The clipped ear club. *ASPCA Animal Watch*, **summer**, 54.
George, W.J. (1974) Domestic cats as predators and factors in winter shortages of raptor prey. *The Wilson Bulletin* **86**, 384-396.
Gibson, K.L., Keizer, K. and Golding, C. (2002) A trap, neuter, and release program for feral cats on Prince Edward Island. *Canadian Veterinary Journal* **43**, 695-698.
Gillies, C.A., Pierce, R., Clout, M. and King, C.M. (2000) Home ranges of introduced mustelids and feral cats at Trounson Kauri Park, New Zealand. *Mammal Review* **303**, 227-232.
Gillies, C.A. and Pierce, R.J. (1999) Secondary poisoning of mammalian predators during possum and rodent control operations at Trounson Kauri Park, Northland, New Zealand. *New Zealand J. Ecology* **23**, 183-192.
Girardet, S.A.B., Veitch, C.R. and Craig, J.L. (2001) Bird and rat numbers on Little Barrier Island, New Zealand, over the period of cat eradication 1976-80. *New Zealand J. Zoology* **28**, 13-29.
Grabarevic, Z., Curic, S., Tustonja, A., Artukovic, B., Simec, Z., Ramadan, K. and Zivicnjak, T. (1999) Incidence and regional distribution of the lungworm *Aelurostrongylus abstrusus* in cats in Croatia. *Veterinarski Archiv* **69**, 279-287.
Gray, F. (1999) Reducing cat predation on wildlife. *Outdoor California* **May-June**, 5-8.
Gunther, I. and Terkel, J. (2002) Regulation of free-roaming cat (*Felis silvestris catus*) populations: a survey of the literature and its application to Israel. *Animal Welfare* **11**, 171-188.
Hall, L.S., Kasparian, M.A., Van Vuren, D. and Kelt, D.A. (2000) Spatial organization and habitat use of feral cats (*Felis catus L.*) in Mediterranean California. *Mammalia* **64**, 19-28.
Harrison, G.H. (1992) Is there a killer in your house? *National Wildlife* **1992**, 10-13.
Haspel, C. and Calhoon, R.E. (1993) The interdependence of humans and free-ranging cats in Brooklyn, New York. *Anthrozoös* **3**, 155-161.
Holton, L. and Manzoor, P. (1993) Managing and controlling feral cat populations: killing the crisis and not the animal. *Veterinary Forum* **March**, 100-101.

Hugh-Jones, M.E., Hubbert, W.T. and Hagstad, H.V. (1995) *Zoonoses: recognition, control, and prevention.* Iowa State University Press, Ames, Iowa.

Hughes, K.L. and Slater, M.R. (2002) Implementation of a feral cat management program on a university campus. *J. Applied Animal Welfare Science* **5**, 15-27.

Hughes, K.L., Slater, M.R. and Haller, L. (2002) The effects of implementing a feral cat spay/neuter program in a Florida county animal control service. *J. Applied Animal Welfare Science* **5**, 285-298.

Huyser, O., Ryan, P.G. and Cooper, J. (2000) Changes in population size, habitat use and breeding biology of lesser sheathbills (*Chionis minor*) at Marion Island: impacts of cats, mice and climate change? *Biological Conservation* **92**, 299-310.

Ishida, Y., Yahara, T., Kasuya, E. and Yanmane, A. (2001) Female control of paternity during copulation: inbreeding avoidance in feral cats. *Behavior* **138**, 235-250.

Izawa, M. (1983) Daily activities of the feral cat *Felis catus* LINN. *J. Mammalogical Society of Japan* **9**, 219-228.

Jackson, J.A. (1978) Alleviating problems of competition, predation, parasitism, and disease in endangered birds. In Temple, S.A. (ed.). *Endangered Birds,* University of Wisconsin Press, Madison, pp. 75-84.

Johnson, K. and Lewellen, L. (1995) *San Diego County: Survey and analysis of the pet population.* San Diego Cat Fanciers, Inc, San Diego, CA.

Johnson, K., Lewellen, L. and Lewellen, J. (1993) *Santa Clara county's pet population.* National Pet Alliance, San Jose, CA.

Kristensen, T. (1980). Feral cat control in Denmark. In *The ecology and control of feral cats.* The Universities Federation for Animal Welfare, Hertfordshire, England, pp. 68-72.

Lawren, B. (1992) Singing the blues for songbirds. *National Wildlife* **Aug/Sept**, 5-11.

Lee, I.T., Levy, J.K., Gorman, S.P., Crawford, P.C. and Slater, M.R. (2002) Prevalence of feline leukemia virus infection and serum antibodies against feline immunodeficiency virus in unowned free-roaming cats. *J. American Veterinary Medical Association* **220**, 620-622.

Levy, J.K., Gale, D.W. and Gale, L.A. (2003a) Evaluation of the effect of a long-term trap-neuter-return and adoption program on a free-roaming cat population. *J. American Veterinary Medical Association* **222**, 42-46.

Levy, J.K., Woods, J.E., Turick, S.L. and Etheridge, D.L. (2003b) Number of unowned free-roaming cats in a college community in the southern United States and characteristics of community residents who feed them. *J. American Veterinary Medical Association* **223**, 202-205.

Liberg, O., Sandell, M., Pontier, D. and Natoli, E. (2000) Density, spatial organisation and reproductive tactics in the domestic cat and other fields. In Turner, D.C. and Bateson, P. (eds.). *The Domestic Cat: the biology of its behaviour,* 2nd edn., Cambridge University Press, Cambridge, pp. 119-147.

Luke, C. (1996) Animal shelter issues. *J. American Veterinary Medical Association* **208**, 524-527.

Macdonald, D.W. and Michael, T. (2001) Alien Carnivores: Unwelcome experiments in ecological theory. *Carnivore Conservation* **5**, 93-122.

Macdonald, D.W., Yamaguchi, N. and Kerby, G. (2000) Group living in the domestic cat: its sociobiology and epidemiology. In Turner, D.C. and Bateson, P. (eds.). *The Domestic Cat: the biology of its behaviour,* 2nd edn., Cambridge University Press, Cambridge, pp. 95-118.

Manning, A.M. and Rowan, A.N. (1998) Companion animal demographics and sterilization status: Results from a survey in four Massachusetts towns. *Anthrozoös* **5**, 192-201.

Martin, G.R., Twigg, L.E. and Robinson, D.J. (1996) Comparison of the diet of feral cats from rural and pastoral Western Australia. *Wildlife Research* **23**, 475-484.

Martinez-Gomez, J.E., Flores-Palacios, A. and Curry, R.L. (2001) Habitat requirements of the Socorro mockingbird Mimodes graysoni. *Ibis* **143**, 456-467.

Mead, C.J. (1982) Ringed birds killed by cats. *Mammal Review* **12**, 183-186.

Meslin, F.X., Fishbein, D.B. and Matter, H.C. (1994) Rationale and prospects for rabies elimination in developing countries. *Current Topics in Microbiology and Immunology* **187**, 1-26.

Murray, R.W. (1992) A new perspective on the problems of unwanted pets. *Australian Veterinary Practitioner* **22**, 88-92.

Mitchell, J.C. and Beck, R.A. (1992) Free-ranging domestic cat predation on native vertebrates in rural and urban Virginia. *Virginia J. Science* **43**, 197-207.

Moller, H. and Alterio, N. (1999) Home range and spatial organisation of stoats (Mustela erminea), ferrets (Mustela furo) and feral house cats (Felis catus) on coastal grasslands, Otago Peninsula, New Zealand: implications for yellow-eyed penguin (Megadyptes antipodes) conservation. *New Zealand J. Zoology* **26**, 165-174.

Muirden, A. (2002) Prevalence of feline leukemia virus and antibodies to feline immunodeficiency virus and feline coronavirus in stray cats sent to an RSPCA hospital. *Veterinary Record* **150**, 621-625.

Natoli, E. (1994) Urban feral cats (*Felis catus L.*): Perspectives for a demographic control respecting the psycho-biological welfare of the species. *Annali Dell'Instituto Superiore di Sanita* **30**, 223-227.

Natoli, E., Ferrari, M., Bolletti, E. and Pontier, D. (1999) Relationship between cat lovers and feral cats in Rome. *Anthrozoös* **12**, 16-23.

Neville, P.F. and Remfry, J. (1984) Effect of neutering on two groups of feral cats. *Veterinary Record* **114**, 447-450.

New, J.C., Jr., Salman, M.D., King, M., Scarlett, J.M., Kass, P.H. and Hutchinson, J.M. (2000) Characteristics of shelter-relinquished animals and their owners compared with animals and their owners in the U.S. pet-owning households. *J. Applied Animal Welfare Science* **3**, 179-201.

Norbury, G. (2000) Predation risks to native fauna following outbreaks of Rabbit Haemorrhagic Disease in New Zealand. *Mammal Review* **30**, 230.

Oliver, E. (2002) Animal welfare in Japan. *Animal People*, **November,** 6.

Olsen, C.W. (1999) Vaccination of cats against emerging and reemerging zoonotic pathogens. *Advances in Veterinary Medicine* **41**, 333-346.

Orloski, K.A. and Lathrop, S.L. (2003) Plague: a veterinary perspective. *J. American Veterinary Medical Association* **222**, 444-448.

Patronek, G.J. (1998) Free-roaming and feral cats-their impact on wildlife and human beings. *J. American Veterinary Medical Association* **212**, 218-226.

Patronek, G.J., Beck, A.M. and Glickman, L.T. (1997) Dynamics of a dog and cat populations in a community. *J. American Veterinary Medical Association* **201**, 637-642.

Read, J. and Bowen, Z. (2001) Population dynamics, diet and aspects of the biology of the feral cats and foxes in arid South Australia. *Wildlife Research* **28**, 195-203.

Remfry, J. (1996) Feral cats in the United Kingdom. *J. American Veterinary Medical Association* **208**, 520-523.

Riordan, A. and Tarlow, M. (1996) Pets and diseases. *British J. Hospital Medicine* **56**, 321-324.

Risbey, D.A., Calver, M.C. and Short, J. (1997) Control of feral cats for nature conservation. I. Field tests of four baiting methods. *Wildlife Research* **24**, 319-326.

Risbey, D.A., Calver, M.C. and Short, J. (1999) The impact of cats and foxes on the small vertebrate fauna of Heirisson Prong, Western Australia I. Exploring potential impact using diet analysis. *Wildlife Research* **26**, 621-630.

Robinson, S.K. (1998) The case of the missing songbirds. *Consequences* **3**, 2-15.

Rosatte, R., Donovan, D., Allan, M., Howes, L.A., Silver, A., Bennett, K., MacInnes, C., Davies, C., Wandeler, A. and Radford, B. (2001) Emergency Response to raccoon rabies introduction into Ontario. *J. Wildlife Diseases* **37**, 265-279.

Schantz, P.M. (1991) Parasitic zoonoses in perspective. *International J. for Parasitology* **21**, 161-170.

Scott, K.C. Levy, J.K. and Crawford, C. (2002) Characteristics of free-roaming cats evaluated in a trap-neuter-return program. *J. American Veterinary Medical Association* **221**, 1136-1138.

Scott, K.C., Levy, J.K. and Gorman, S.P. (2002) Body condition of feral cats and the effect of neutering. *J. Applied Animal Welfare Science* **5**, 203-213.

Seabrook, W. (1989) Feral cats (Felis catus) as predator of hatchling green turtles (Chelonia mydas). *J. Zoology* **219**, 83-88.

Serpell, J. A. (2000) Domestication and history of the cat. In Turner, D.C. and Bateson, P. (eds.). *The Domestic Cat: the biology of its behaviour*, 2nd edn., Cambridge University Press, Cambridge, pp. 179-192

Shah, N.J. (2001) Eradication of alien predators in the Seychelles: an example of conservation action on tropical island. *Biodiversity and Conservation* **10**, 1219-1220.

Short, J., Turner, B., Risbey, D.A., Danielle, A. and Carnamah, R. (1997) Control of feral cats for nature conservation. II. Population reduction by poisoning. *Wildlife Research* **24**, 703-714.

Slater, M.R. (2000) Understanding and controlling of feral cat populations. In August, JR (ed.), *Consultations in feline internal medicine*, 4th edn., W.B. Saunders, Philadelphia, pp. 561-570.

Slater, M.R. (2002) *Community approaches to feral cats: problems, alternatives & recommendations*. The Humane Society Press, Washington, D.C.

Slater, M.R. (2003) Current concepts in free-roaming cat control. *Proceedings of the 10th International Symposium of Veterinary Epidemiology and Economics*, Vina del Mar, Chile.

Smith, R.E. and Shane, S.M. (1986) The potential for the control of feral cat populations by neutering. *Feline Practice* **16**, 21-23.

Sorace, A. (2002) High density of bird and pest species in urban habitats and the role of predator abundance. *Ornis Fennica* **79**, 60-71.

Tabor, R. (1983) *The wild life of the domestic cat*. Arrow Books Limited, London.

Tan, J.S. (1997) Human zoonotic infections transmitted by dogs and cats. *Archives of Internal Medicine* **157**, 1933-1943.

Terborgh, J. (1992) Why American songbirds are vanishing. *Scientific American* **May**, 98-104.

Turner, D. C. (2000) The human-cat relationship. In Turner, D.C. and Bateson, P. (eds.). *The Domestic Cat: the biology of its behaviour*, 2nd edn., Cambridge University Press, Cambridge, pp. 193-206.

Twyford, K.L., Humphrey, P.G., Nunn, R.P. and Willoughby, L. (2000) Eradication of feral cats (Felis catus) from Gabo Island, south-east Victoria. *Ecological Management and Restoration* **1**, 42-49.

Veitch, C.R. (2001) The eradication of feral cats (Felis catus) from Little Barrier Island, New Zealand. *New Zealand J. Zoology* **28**, 1-22.

Wenstrup, J. and Dowidchuck, A. (1999) Pet overpopulation: data and measurement issues in shelters. *J. Applied Animal Welfare Science* **2**, 303-319.
WHO (2002a) Rabies vaccines: WHO position paper. *Weekly Epidemiological Record* **77**, 109-119.
WHO (2002b) World survey of Rabies for the year 1999. **35**, www.who.int/emc-documents/rabies/whocdscsreph200210.html.
WHO Expert Committee (1988) Report of WHO consultation on dog ecology studies related to rabies control. *World Health Organization* **88.26**, 1-35.
WHO Expert Committee (1994) Report of the fifth consultation on oral immunization of dogs against rabies. *World Health Organization* **94.45**, 1-24.
Williams, L.S., Levy, J.K., Robertson, S.A., Cistola, A.M. and Centonze, L.A. (2002) Use of the anesthetic combination of tiletamine, zolazepam, ketamine, and xylazine for neutering feral cats. *J. American Veterinary Medical Association* **220**, 1491-1498.
Work, T.M., Massey, J.G., Rideout, B.A., Gardiner, C.H., Ledig, D.B., Kwok, C.H. and Dubey, J.P. (2000) Fatal toxoplasmosis in free-ranging endangered 'Alala from Hawaii. *J. Wildlife Diseases* **36**, 205-212.
Yamane, A., Emoto, J., and Ota, N. (1997) Factors affecting feeding order and social tolerance to kittens in the group-living feral cat (*Felis catus*). *Applied Animal Behaviour Science* **52**, 119-127.
Yilmaz, H., Ilgaz, A. and Harbour, D.A. (2000) Prevalence of FIV and FeLV infections in cats in Istanbul. *J. Forensic Sciences* **2**, 69-70.
Zasloff, R.L. and Hart, L.A. (1998) Attitudes and care practices of cat caretakers in Hawaii. *Anthrozoös* **11**, 242-248.
Zaunbrecher, K.I. and Smith, R.E. (1993) Neutering of feral cats as an alternative to eradication programs. *J. American Veterinary Medical Association* **203(3)**, 449-452.
Zawistowski, S., Morris, J., Salman, M.D. and Ruch-Gallie, R. (1998) Population dynamics, overpopulation and the welfare of companion animals: new insights on old and new data. *J. Applied Animal Welfare Science* **1**, 193-206.

Chapter 7

HOUSING AND WELFARE

Irene Rochlitz
Animal Welfare and Human-animal Interactions Group, Department of Veterinary Medicine, University of Cambridge, Madingley Road, Cambridge CB3 0ES, UK

Abstract: The way a cat is housed will have a significant impact on its welfare. The range of housing conditions in which cats may be kept include boarding, breeding and quarantine catteries, shelters, research facilities, veterinary practices and the home. Drawing on ethological principles, the evolutionary history of the cat and studies of cats kept in different conditions, the housing requirements of cats are described with regard to the quantity and quality of space, contact with conspecifics and with humans, and the sensory, occupational and nutritional environments. Features specific to research facilities, shelters and the home are considered, as are the advantages and disadvantages of confining pet cats indoors or allowing them outdoor access.

1. INTRODUCTION

A number of publications on cat behaviour, welfare, human-cat interactions, and other cat-related topics, including housing, have emerged in the last few decades, though research in this area remains limited. Most studies about cat housing have been conducted in laboratories, boarding and quarantine catteries and shelters (e.g. McCune 1995; Rochlitz et al. 1998; Ottway & Hawkins 2003). A few publications make recommendations about the enrichment of the home environment for cats confined indoors, with the aim of preventing behavioural problems (e.g. Schroll 2002). Others describe aspects of cat behaviour (Bernstein & Strack 1996; Barry & Crowell-Davis 1999), and interactions between humans and cats (Mertens & Turner 1988; Mertens 1991) in the home setting. This chapter aims to bring together research findings that contribute to our understanding of how cats should be housed and cared for in order to maximize their welfare.

It is obvious that the way a cat is housed and looked after will have a profound influence on its welfare (see Chapter 2). It may initially seem particularly important that housing should be good where cats spend their entire life confined there, such as pet cats kept indoors or cats in laboratories. However, whether the cat will be housed in a particular environment for two days (for example, in a veterinary hospital), for two weeks (in a boarding cattery), two months (in a shelter) or two years (in a laboratory) is of little relevance to the animal. Its welfare is largely determined by the conditions it lives in day-by-day, so high standards of housing and care should apply to all the situations in which cats are kept.

This chapter first describes the main types of housing used for cats, and then considers their most important characteristics with regard to quantity of space; quality of space; and the social, sensory, occupational and nutritional environments. Features of three specific housing conditions, research facilities, shelters and catteries, and the home environment (including whether cats should be confined indoors or allowed outdoor access), are presented.

2. MAIN HOUSING CONDITIONS

The main housing conditions in which cats may be kept are shown in Table 1. In many countries, research facilities and most catteries are required to meet certain minimum standards and are therefore licensed and inspected on a regular basis, while other types of housing, such as shelters, are exempt from regulation. The need to license animal shelters and sanctuaries has been recognised (Patronek & Sperry 2001; Companion Animal Welfare Council 2004). Replacing the Protection of Animals Act 1911 with an Animal Welfare Bill, which would allow the creation of a new offence of 'likely to cause unnecessary suffering', is currently under discussion in the United Kingdom (Department for Environment, Food and Rural Affairs 2002). Under this bill, owners of companion animals would be required to provide a minimum standard of care, which includes adequate housing conditions (a similar bill has already been enacted in Queensland, Australia). The introduction of a Practice Standards Scheme is also planned in the United Kingdom, and will require veterinary practices to provide a minimum standard of accommodation for their hospitalised patients (Royal College of Veterinary Surgeons 2004).

Table 1. The main housing situations in which cats may be kept, the principal caregiver and whether regulatory legislation exists in the United Kingdom.

Housing situation	Principal caregiver	Regulatory legislation (United Kingdom)
research facility	facility staff	yes
boarding or quarantine cattery	cattery staff	yes
breeding cattery	owner and/or cattery staff	yes (if large scale)
shelter or sanctuary	shelter staff and/or volunteers	no
veterinary practice	veterinarian and/or nurse	no
home	owner	no

3. HOUSING REQUIREMENTS

3.1 General Recommendations

Traditionally, much of the advice on cat housing has been based on what is generally practised and what is most convenient for caretakers. Current recommendations can draw on this traditional approach, as well as on recent research (which largely relies on observational studies) and advice from those experienced in the field, such as ethologists and animal behaviour counsellors. More research is needed on the requirements of cats in different housing conditions, so it is likely that recommendations will be modified as further knowledge is gained.

An important objective of good housing is to improve welfare by giving the animal a degree of control over its environment (Broom & Johnson 1993). While housing in a barren environment leads to apathy and boredom (Wemelsfelder 1991), cats do not like unpredictability such as irregular contact with unfamiliar cats or humans, or an unfamiliar and unpredictable routine (Carlstead *et al.* 1993). How the cat responds to the level of stimuli, or predictability, in the environment will depend on many factors, including the cat's temperament (Lowe & Bradshaw 2001) and previous experiences. Providing extremes are avoided, a cat that has a variety of behavioural choices and is able to exert some control over its physical and social environment will develop more flexible and effective strategies for coping with stimuli.

Cats are more likely to respond to poor housing conditions by becoming inactive and by inhibiting normal behaviours such as self-maintenance (feeding, grooming and elimination), exploration and play, than by overtly showing abnormal behaviour (see Chapter 2). Sick cats will modify their behaviour in a similar way (see Chapter 8). Keeping cats in an environment that encourages a wide range of normal behaviours will, therefore, make it easier for caretakers or owners to detect when cats are unwell.

The control of infectious disease is very important, particularly when cats are housed in groups and where the composition of the group is unstable (see Chapter 8). Care should be taken that management and environmental enrichment procedures do not increase the risk of disease transmission. Over-emphasis on the need for sanitary conditions, however, can lead to a barren housing environment.

3.2 Quantity of Space

Barry and Crowell-Davis (1999) examined gender differences in the social behaviour of the neutered indoor-only cat. While this study did not set out to establish the amount of space cats require, indirectly it gives us some clues. Sixty pairs of cats were studied: 20 were male-male pairs, 20 were female-female and 20 were female-male pairs (47 of the 60 pairs were not related). All the cats were neutered, were never allowed outside, were between 6 months and 8 years of age, and had lived together for at least 3 months. While the cats spent half of their time out of each other's sight, for most of the time that they were together (25 to 31% of observed time) they kept a distance between themselves of 1 to 3 metres. Intriguingly, the male-male pairs spent more time in close proximity (0 to 1 metre; 19% of observed time) than female-female (8.8% of observed time) and female-male (8.6% of observed time) pairs. All cats spent more than 35% of the time within 3 metres of each other. Gender had little effect on the cat's social behaviour, and there were low levels of aggression; the cats regularly ate together and shared resting areas. A reasonable conclusion from this study is that when cats are together, there should be enough space so that they can maintain distances between themselves of at least 1 metre (this can include vertical distance).

Other studies have attempted to determine the minimum size of enclosure (the term enclosure refers to a cage or pen in a cattery, animal shelter or laboratory as well as to the home environment) that cats need, particularly in situations where space is at a premium. Kessler and Turner (1999b) suggest that there should be at least 1.7 m^2 per cat for group-housed cats in shelters (see section 4.2.1). The working party for the review of the European Convention for the Protection of Vertebrate Animals used for Experimental and Other Scientific Purposes (ETS 123), Appendix A, Council of Europe (1986), recommends that one cat can be housed in a cage with a minimum floor area of 1.5 m^2, with another 0.75 m^2 for every additional cat; the cage should be high enough for humans to enter (walk-in) (R. Hubrecht, personal communication).

Domestic cats, having evolved from the semi-arboreal African wild cat, spend less time on the floor of their pens than on raised surfaces (Podberscek et al. 1991; Rochlitz et al. 1998), and high structures, which provide vantage points, are used more frequently than low ones (Durman 1991; Roy 1992; Smith et al. 1994). As the vertical dimension is so important for cats enclosures should be of adequate height, at least 1.5 m so that the cat can stretch fully and jump freely. Walk-in enclosures are ideal, as they also allow caretakers to enter and interact closely and comfortably with the cats. If an enclosure is too small, there may be an increase in agonistic encounters or cats will attempt to avoid each other by decreasing their activity (Leyhausen 1979; van den Bos & de Cock Buning 1994a).

In some instances, it may be necessary to house cats singly in small cages, for example when they are recovering from an experimental procedure or are hospitalised in a veterinary practice. The cage should have at least 1.5 m^2 of floor space, and ideally should be no less than 1 m high and contain at least one shelf (unless this is contraindicated by the procedure), which will allow the cat to rest on an elevated surface and still be able to stretch in the vertical direction. Placing the cage on a shelf at waist height or higher will make access easier for the caretaker. A litter tray should be provided, as well as a semi-enclosed retreat area such as a box or deep-sided tray (see next section).

3.3   Quality of Space

Beyond a certain minimum size of enclosure, it is the quality rather than the quantity of space that is most important. Most cats enjoy climbing and jumping and, as mentioned previously, spend much of their time off the floor; they use elevated areas as vantage points from which to monitor their surroundings (DeLuca & Kranda 1992; Holmes 1993; James 1995). There should, therefore, be structures within the enclosure that enable cats to use the vertical dimension, such as shelves, climbing posts, walkways, windowsills and platforms.

Cats spend a large portion of their day either resting or sleeping, so it is important that the rest areas have comfortable surfaces (Figure 1). One study noted an improvement in welfare when laboratory cats were provided with soft resting surfaces in the form of pillows (Crouse et al. 1995), while another found that cats preferred polyester fleece to cotton-looped towel, woven rush-matting and corrugated cardboard for lying on (Hawthorne et al. 1995). In a study of environmental enrichment of cats in rescue centers, Roy (1992) found that cats preferred wood as a substrate to plastic, and also liked materials that maintain a constant temperature such as straw, hay, wood shavings and fabric. As cats are more likely to rest alone than with others

(Podberscek *et al.* 1991), there should be a sufficient number of comfortable resting areas for all cats in the enclosure.

Figure 1. Cats should have comfortable resting areas. This cat is lying in a hammock, which is hung over a radiator. (Courtesy of Cerian Webb).

Hiding is a coping behaviour that cats often show in response to stimuli or changes in their environment (see Chapters 2 and 4). It is commonly seen when cats want to avoid interactions with other cats or people, and in response to other potentially stressful situations (Carlstead *et al.* 1993; James 1995; Rochlitz *et al.* 1998). As mentioned previously, the study by Barry and Crowell-Davis (1999) of 60 pairs of neutered, indoor-only cats, found that they spent between 48 and 50% of the observed time out of each other's sight. Therefore, in addition to open resting areas (such as shelves) there should be resting areas where cats can retreat to and be concealed, such as high-sided cat beds, 'igloo' beds and boxes. Visual barriers such as vertical panels, curtains and other room divisions, can also be useful to enable cats to get out of sight of others. Vertical room dividers will also break up the space into compartments, making it more complex and giving the cat more choice about where it wants to be.

There should be a sufficient number of litter trays, at least one per two cats and preferably one per cat, sited away from feeding and resting areas. Cats can have individual preferences for litter and tray characteristics, so it may be necessary to provide a range of litter types and designs of litter trays (covered or open) (see Chapter 4).

3.4 The Social Environment

3.4.1 Conspecifics

The cat is a social carnivore that regularly interacts with conspecifics (Leyhausen 1979). In research facilities, multi-cat households and some animal sanctuaries, cats are expected to spend most of their lives together. The majority of cats can be housed in groups providing that they are well socialized to other cats, and that there is sufficient good-quality space, easy access to feeding and elimination areas and a sufficient number of concealed retreats and resting places. Ideally, the composition of the group should be stable, with minimal additions or losses of cats (see section 4.3.2 for a discussion of environmental stressors due to social factors, and Chapters 1, 2 and 4 on socialization). Many factors will determine the ideal group size, but it seems that 10 to 12 for shelters and 20 to 25 for cats in laboratories is an appropriate maximal number (James 1995; Hubrecht & Turner 1998). Cats that fail to adapt satisfactorily to living in groups should be identified and housed in pairs or singly.

When there are many cats housed together, it may be necessary to distribute feed, rest and elimination areas in a number of different sites to prevent certain cats from monopolising one area and denying others access (van den Bos & de Cock Buning 1994b). Conflict between cats may arise, for example, if feeding bowls are concentrated in a small area or all the litter trays are placed together.

Neutered cats can be kept together in groups, as can entire females. While some authors suggest that entire males should be housed singly, others have shown that they can be housed successfully with other entire males (Hart 1980) and with neutered males (Podberscek *et al.* 1991); they can also be kept with neutered females.

3.4.2 Humans

The caregiver, whether a member of shelter staff, an animal technician, an owner or a veterinary nurse, is the most important determinant of the cat's welfare, and the best housing conditions cannot substitute for the caregiver's compassionate care and attention. Unfortunately, it is often this direct involvement of staff in animal care that is curtailed, especially in situations where there are limits on financial resources, shortage of time or high work demands. While interactions with conspecifics or other animals are also important and rewarding to the cat, they are not a substitute for human attention. Randall *et al.* (1990) found that laboratory cats organized their daily activity patterns around human caregiver activity, and responded strongly to humans in their environment. Cats in enriched conditions in a laboratory facility demonstrated a clear preference for human contact over toys (DeLuca & Kranda 1992). Periods of time, which are not part of routine care-taking procedures (such as feeding or cleaning), should be available every day for cats to interact with their caregiver (Figure 2). Some cats may prefer to be petted, groomed and handled while others may prefer to interact via a toy (Karsh & Turner 1988).

Figure 2. A period of time, which is not part of routine care-taking procedures, should be set aside every day for cats to interact with their care-giver.

In order to care for them properly, the caregiver should like cats and be knowledgeable about them. This knowledge can be acquired from many sources, such as books and other printed information, internet sites, veterinary practices, animal behaviour courses, animal behaviourists, animal rescue charities, and from mentoring and supervision by peers. Because some information may be controversial, contradictory or wrong, inexperienced caregivers will require guidance from reputable sources.

In situations where cats are kept as companion animals in the home, the benefits to humans from caring properly for their pet are obvious (see Chapter 3). Benefits can also arise in other, less obvious conditions, such as in a shelter or laboratory. Allowing caregivers to enrich the environment of cats under their care can also enrich their own lives (Young 2003). This involvement introduces variation in their work, provides them with opportunities to learn about the species, to devise the enrichment and to observe its effects, and, by improving the cats' welfare, makes their work more rewarding.

3.5 The Sensory Environment

The quality of the external environment is very important to cats, whose senses are highly developed (Bradshaw 1992). Cats spend a lot of time observing the environment immediately outside their enclosure; they will often settle on windowsills if they are wide and comfortable enough, but other suitable vantage points, such as climbing platforms and shelves placed near windows, may be used. DeLuca and Kranda (1992) found that research cats housed as a group in a room spent most of the day sitting on a window perch, watching activity in the outside hallway. If cats do not have free access to the outdoors they should have access to enclosed outdoor runs or, if this is not possible, their enclosure should have windows so that they can look outside (Figure 3).

Olfactory enrichment is relatively underused in animal housing, perhaps because of the relatively poor sense of smell of humans compared with many other species. Recently, Wells and Egli (2003) examined the effect of introducing four odours on the behaviour of six zoo-housed black-footed cats (*Felis nigripes*). The odours were nutmeg, catnip (*Nepeta cataria*), body odour of prey (quail), and no artificial odour (as a control) and were introduced individually into the cats' environment on impregnated flannel cloths. There was an increase in the amount of time cats spent in active behaviours and a decrease in time spent in sedentary behaviours, though the response to the odours waned over the 5-day observation period. Nutmeg caused less of an effect than catnip or odour of prey. Catnip is well known as a stimulant for cats, though not all cats are affected by it. It is usually supplied as a dried herb or in toys, and can be grown as a fresh plant in pots.

Cats have an excellent sense of smell and olfactory communication is important in this species (Bradshaw & Cameron-Beaumont 2000), although the role that odours play within social groups is not well understood (Bradshaw 1992; see Chapter 1). Sebaceous glands are located throughout the body, especially on the head and the peri-anal area, and between the digits. Scratching, which causes scent to be deposited from the inter-digital glands, is frequently observed in cats; this marking behaviour also leaves visual signals (striations) and helps to maintain the claws in good shape (see Chapter 4). Surfaces for the deposition of these olfactory and visual signals and for claw abrasion, such as scratch posts, rush matting, pieces of carpet and wood, should be provided.

Figure 3. Cats should have access to natural light, and be able to observe the outside environment. (Courtesy of Cerian Webb).

In some animal houses auditory enrichment using a radio, to provide music and human conversation, is thought to prevent animals from being startled by sudden noises and habituate them to human voices, and to provide a degree of continuity in the environment (Benn 1995; James 1995; Newberry 1995). Video recordings of images and sounds that are thought to appeal to cats are available, though their effectiveness has not been evaluated.

3.6 The Occupational Environment

Many cats play alone or with their owners, rather than with other cats (Podberscek *et al.* 1991), so there should be enough space for them to play without disturbing others. A variety of toys should be available, and they should be replaced regularly as novelty is important to cats. Many toys are made to resemble mice and other small animals and to look attractive to humans, but they are often not very effective at eliciting play or pseudo-predatory behaviour in the cat. The qualities of the toy are more important than its appearance: it does not matter if it looks artificial, providing it elicits the desired behaviour. Objects which are mobile, have complex surface textures and mimic prey characteristics are the most successful at promoting play (Hall & Bradshaw 1998). For laboratory cats caged singly, de Monte and Le Pape (1997) found that a tennis ball was a more effective enrichment tool than a wooden log.

Cats also benefit from opportunities to explore, so suitable novel objects such as boxes, large paper bags and other structures can be introduced into their environment intermittently.

3.7 The Nutritional Environment

Domestic cats are usually offered two or three meals a day and seem to adapt well to this, although their preferred pattern of feeding is one of frequent small meals (Bradshaw & Thorne 1992; see Chapter 9). Frequent feeding may not always be possible in the home environment, and *ad libitum* feeding may lead to obesity.

Another environmental enrichment technique is to increase the time animals spend in pseudo-predatory and feeding behaviour. Studies have examined the effects of food presentation in a number of species, including captive small wild cats (Markowitz & LaForse 1987; Law *et al.* 1991; Shepherdson *et al.* 1993). McCune (1995) suggests putting dry food into containers with holes through which the cat has to extract individual pieces. For the cat that is food-orientated, small amounts of dry food can be hidden in the environment to make it more interesting to explore. Toy-like objects that are destructible and have nutritional value may be of interest to cats, but there are few such items available commercially.

Cats often prefer to drink away from the feeding area, so bowls of water should be placed in a number of locations, both indoors and outdoors. Schroll (2002) states that cats like to be in a slightly downward position when drinking, and some like drinking running water, such as from a dripping tap or small water fountain.

Grass grown in containers can be provided for indoor-only cats; some cats like to chew it and it is thought that this can help with the elimination of furballs (trichobezoars) (Figure 4).

Figure 4. Grass can be grown in pots for cats confined indoors; some cats like to chew it and it is thought that this can help with the elimination of furballs.

4. SPECIFIC HOUSING CONDITIONS

4.1 Research Facilities

Whenever animals are to be used in biomedical research, consideration should be given to the implementation of the 'Three Rs': replacement, reduction and refinement (Russell & Birch 1959). While the ultimate aim should be to replace all live animal use in experiments with non-sentient material, it is likely that cats will continue to be used in such research in the near future, albeit in reducing numbers, and refinement remains very important. Refinement applies both to experimental procedures and to the way cats are housed and looked after. While much attention, justifiably, is paid to the regulation of experimental procedures, with the emphasis on the control of pain, housing conditions also have a major impact on the cats' welfare so they too should be well regulated to the highest standard (Figure 5). Keeping cats in an enriched, stimulating environment that encourages a wide range of normal behaviours will, by enhancing their welfare, make them better subjects for scientific investigation (Poole 1997), have a positive effect on the public perception of the treatment of animals in laboratories (Benn 1995) and, when these cats are no longer required for research and are re-homed, they will be more likely to adapt successfully to their new home environment.

Figure 5. There are toys, shelves, a scratch post and access to an outdoor pen (via a cat-flap) in this research facility.

It has been suggested that the emphasis in laboratory animal housing should be shifted from an 'engineering' approach (providing cages of certain dimensions and features, and defined management procedures) to a 'performance' approach (providing housing conditions and management procedures that enable the animals to reach certain performance standards) (National Research Council 1996). While the performance approach is more flexible and less prescriptive, the engineering approach is sometimes useful to establish a baseline. Because cage space in laboratories is costly, a number of studies have examined the effect on animals of enlarging cage size (see Reinhardt & Reinhardt (2001) for a summary). Assessing only the effect of the quantity of cage space on the behaviour of animals is of limited value, because the usefulness of the space depends mostly on the quality of the cage contents that enable the animal to make use of the additional space. A small cage that is barren will still be barren if it is a little bigger. Using the engineering approach, a minimum cage size that is able to contain the basic features which are necessary to make the quality of the space appropriate for cats can be specified, but thereafter the performance approach can be used to develop more imaginative and flexible ways of enriching the cage.

Because of security concerns, the need to control costs and other factors, the trend in housing cats in research facilities has been towards keeping them in rooms, often without windows, within buildings rather than in enclosures with outdoor runs. Depriving cats of sensory access to the external environment around their enclosure is likely to have a detrimental effect on their welfare.

4.2 Shelters and Catteries

The Chartered Institute of Environmental Health has published model licence conditions and guidance for cat boarding establishments in the United Kingdom (CIEH Animal Boarding Establishments Working Party 1995), which serve as a basis upon which local environmental health officers issue licences to boarding catteries. The Feline Advisory Bureau (a cat charity) in the United Kingdom has published two manuals, one on how to set up and manage a boarding cattery (Bessant 2002) and another on how to set up and manage a shelter (Haughie 1998). While much of the advice is sound and based on experience and current practice, there is a need for more scientific input into the best way to house cats in catteries and shelters. Although more studies have been carried out in shelters than in boarding catteries, research findings can, in most instances, be extrapolated from one environment to the other.

4.2.1 Role of Shelters

The function of shelters is to provide housing, food and care for cats that are abandoned and unwanted and, providing the cats are healthy, to find them homes as quickly as possible. A cat's stay in the shelter should be kept short and the cat subjected to as little stress as possible. The population of cats entering shelters is often extremely heterogeneous (Evans 2001), differing, for example, in origin (feral, stray, owned), socialization status, age, vaccination status and health. In most shelters, the control of infectious disease is a major challenge (see Chapter 8). In a study of respiratory and enteric viruses in 162 cats entering shelters in the United States, feline calicivirus (FCV) was isolated from 11% of healthy cats upon entry, feline herpesvirus (FHV) from 4% and feline enteric coronavirus (FECV) from 33% of cats (Pedersen *et al.* 2004). The subsequent spread of all three viruses was rapid: 15% of cats were shedding FCV, 52% FHV and 60% FECV after one week. Although the shelter environment may not be the primary source of these viruses, it serves to spread viruses between infected and non-infected cats, to reactivate latent infections and to enhance the severity of disease through stress and increased exposure.

The Cat-Stress-Score (CSS), which is based on body movements, postures and other features such as pupil size, has been widely used in studies of cats in shelters and catteries (see Chapter 2). It is most useful for assessing the stress levels of cats confined to a small space such as a cage or pen; changes in scores for an individual cat, taken over a period of time, can reflect how it is adapting to the new environment.

Cats with previous experience of boarding in catteries or shelter-like accommodation, as well as those that have short traveling times to the premises and short waiting times before being admitted, will settle in more quickly and be less stressed (McCune 1992). While Kessler and Turner (1997) did not find that the age of the cats affected their adjustment to being housed in shelters and boarding catteries, in McCune's study (1992) the stress levels of older cats declined more rapidly. Kessler and Turner (1999b) examined how the density at which cats were housed in groups in a shelter affected their stress levels. There was a positive correlation between CSS and group density, and a minimum floor space of 1.7 m^2 per cat was recommended to ensure acceptable stress levels. However, these cats were socialized to conspecifics, the composition of the groups was relatively stable, and the enclosures well adapted for cats. In other situations, more space per cat may be required.

Some cats may be housed in shelters for long periods of time (months or even years), especially if the shelter has a 'no-kill' policy, that is they will not euthanise a healthy animal. Due to the social disruption, lack of control, and both acute and chronic fear-inducing situations that may exist in the shelter environment, concerns about the welfare of these long-stay animals have been raised (Patronek & Sperry 2001). Ensuring that the cat has daily, rewarding contact with humans is important, and may have other beneficial effects. Hoskins (1995) examined the effect of human contact on the reactions of cats in a rescue shelter: cats that received additional handling sessions, where they interacted closely with a familiar person, could subsequently be held for longer by an unfamiliar person than cats that did not receive additional handling sessions. This is likely to improve their rehoming potential. Siegford *et al.* (2004) describes a behavioural test that could be used to better match cats with prospective owners (see Chapter 3).

4.2.2 Single versus Group Housing

There has been much discussion and some research on whether cats should be housed discretely or in communal groups in shelters.

Ottway and Hawkins (2003) studied 72 cats living long-term (over one month) in shelters. Thirty-six were housed communally with non-familiar conspecifics and 36 in discrete units, either singly or with another one or two previously familiar conspecifics. The mean CSS was higher in communal than in discrete-unit housing, though very high stress scores (more than 5 out of 7) were not recorded in cats housed under either condition. Cats housed communally spent more time hidden, while cats in discrete housing were more likely to play and spent more time resting or sleeping in contact with another cat. They concluded that communal housing may be undesirable in a shelter situation, because of the inappropriate social grouping of unrelated adult cats and inherent instability of the group.

Durman (1991) studied the behaviour of cats housed communally in small groups (between four and seven cats) in rooms in a shelter. Newly introduced cats were aggressive towards others, and showed behaviours indicative of high levels of stress (such as vocalizing and attempting to escape). These behaviours had largely disappeared after four days while other behavioural measures (such as sitting underneath a shelf, exploring the room and sitting alertly) changed more slowly, but all had reached equilibrium after two weeks. Some cats, who had been at the shelter for more than a few months, had access to large outdoor pens. The most recently-introduced cats to the group were the most vigilant and aggressive, while those present in the shelter for more than one year were more likely to rest in contact with another cat, to approach other cats, and to initiate rubbing and mutual grooming. While Smith *et al.* (1994) did not find behaviour patterns indicative of high stress levels in a group of unrelated cats in a shelter, these cats were part of a stable group living together long-term and there was not a constant influx of new cats.

Gourkow (2001) examined the effects of four different housing conditions, or treatments, on the behaviour and stress levels of cats in a shelter, and on the outcome for these cats (whether they were adopted, became ill or were euthanised). The first treatment, T1, consisted of a small stainless-steel cage containing food and water bowls, a litter tray and a folded towel as bedding. The second treatment, T2, was the same cage enriched with a wooden shelf and hiding area underneath it. In these first two treatments cats were housed singly. The third treatment, T3, was a walk-in cage converted from a dog kennel, where cats were housed in groups of eight. The cage had ten shelves, five hiding areas, bedding and a small chair, as well as food and water bowls and litter trays. The fourth treatment, T4, was the same as T3 except that it contained more furniture items, such as a plastic cat playhouse, toys and a scratching post. Cats in T1 were handled by a number of caretakers, each of whom handled the cat in their own way, whereas cats in the other treatments were handled by only one or two people and in a more consistent manner.

Cats in T1 had higher stress scores, were less likely to be adopted, and were adopted at a later stage than cats in the other treatments. They were also more likely to be euthanised. Compared to cats in the other treatments, T1 cats were less likely to display behaviours that potential adopters described as desirable (such as interacting with another cat). While there were a number of variables that differed between treatment groups, and these were not controlled for in comparisons, this study illustrates the effects of housing and of contact with humans on the welfare of cats, both in terms of the levels of stress they experienced while in the shelter and on their ultimate fate.

Cats entering a shelter have to cope with the stress of leaving their own familiar environment (in the case of owned cats), of entering a strange new environment, of being handled by strangers and, in most cases, undergoing a veterinary examination. They will be aware that there are other unfamiliar animals nearby. In the first few weeks following arrival, it is unlikely that they will benefit from being housed with unfamiliar conspecifics or in a group of cats whose composition is constantly changing. During this period, it is probably preferable to house cats in discrete units, that is to keep them in their original groups (four or more cats from the same household can be split into smaller groups of two to three cats), rather than introduce incoming cats into groups of cats with whom they are not familiar (Figure 6). This period of discrete housing will also allow caretakers to find out more about the individual cat's health, behaviour and personality, and to identify, treat and control disease. If adoption is not imminent or if there are constraints on space, it may then be worth considering moving the cat into communal housing providing that the group is not too large, there is plenty of space that is suitably enriched, and that there is some stability in group composition. Cats previously socialized toward people and conspecifics will adapt better to housing in groups than non-socialized cats (Kessler & Turner 1999a). There will be some cats that are unable to adapt to communal housing; they should be identified and housed in pairs if possible, or singly.

Figure 6. Cats entering this shelter are housed in their original groups, and not mixed with cats with whom they are not familiar.

4.3　Home Environment

4.3.1　Indoor-only versus Outdoor Access

In the United Kingdom, the majority of cats are allowed access to the outdoors; it is generally considered that this is the natural thing for cats to be able to do. In a questionnaire survey of owners of 1,070 cats, 90 cats (8.4%) were confined indoors; 70% were under a year of age and it was likely that most of the young, entire animals would eventually be allowed out once neutered (I. Rochlitz, unpublished data). In the United States, between 50 and 60 per cent of pet cats are kept permanently indoors (Patronek *et al.* 1997). The American Veterinary Medical Association has stated that it strongly encourages owners of domestic cats in urban and suburban areas to keep them confined indoors; the Humane Society of the United States and many American shelters also commonly advise this, as do American veterinarians (Buffington 2002). In many areas of Australia, concerns about the effects of cat predation on wildlife have led to the adoption of regulations restricting pet cats' access to the outdoors. In addition, the population of Australian cats is declining due to neutering (de-sexing) being compulsory in many areas (C. Phillips, personal communication).

It is generally assumed that cats confined indoors will be healthier and live longer, as they are protected from hazards associated with the outdoors (Table 2). Different hazards, however, may be present in the home. The Blue Cross animal charity found that household accidents were the second most common reason that puppies and kittens were brought into its flagship hospital (Veterinary Department, The Blue Cross, London, personal communication). The accidents included falls from balconies and windows, kitchen scalds and burns, and access to cleaning products.

Buffington (2002) recently reviewed the veterinary literature for epidemiological data on cats confined indoors and disease risk (Table 2). He found that conditions such as feline urologic syndrome (a urinary tract disease), odontoclastic resorptive lesions (a dental disease), obesity, hyperthyroidism (an endocrine disease), and behavioural problems (such as inappropriate elimination) were associated with keeping cats indoors in some studies, while others did not find that indoor cats were at increased risk of developing these conditions. The difficulty with conducting these studies is that there are likely to be confounding factors and interactions. For example, pedigree cats may be more likely to be kept indoors, may be predisposed to certain inherited diseases, and may be treated differently by their owners compared with non-pedigree cats.

Table 2. Conditions and diseases that may be associated with confining a cat indoors or allowing it access to the outdoors

Cat is confined indoors	Cat has access to the outdoors
feline urologic syndrome	infectious diseases (e.g. viral, parasitic)
odontoclastic resorptive lesions	road traffic accidents
hyperthyroidism	other accidents (e.g. falling from a tree)
obesity	fights with other cats
household hazards	attacks by humans, dogs and other animals
behavioural problems (e.g. inappropriate elimination/toileting)	poisoning
boredom	theft
inactivity	going astray

While a pet cat with outdoor access can probably compensate to some degree for poor conditions in the home (Turner 1995b), eventually the cat may leave the home to find better conditions elsewhere, become a stray and end up in a shelter. Only a small proportion of cats are identified with a microchip or collar, and the majority of stray cats entering shelters are not reclaimed (Evans 2001; see Chapters 5 and 6). Owners of lost cats often find it very difficult to trace them, especially if the cats do not have a form of identification.

The main concern with an indoor environment is that, compared with the outdoors, it is relatively impoverished, predictable and monotonous and may cause the cat to experience boredom and stress. While we often do not know what cats do when they are outdoors, it is generally assumed that indoor cats are less active and that this inactivity can lead to obesity and other problems. Certain behaviours, such as scratching items and spraying urine, may be considered normal when performed by a cat outdoors but become problematic when performed indoors (see Chapter 4).

As discussed previously, most cats seem to be able to adapt to indoor living well providing they have been kept in this kind of environment from an early age. Some geriatric cats, or those with disabilities, may also benefit from being confined indoors but cats used to having outdoor access may have difficulty adapting to an entirely indoor existence when adult (Hubrecht & Turner 1998). The recent increase in popularity of the cat as a companion animal in many countries has been partly ascribed to the fact that it requires relatively little care compared with the dog, not having to be taken out for walks or be trained. Also, it can be kept in a smaller space, such as an apartment, and will use a litter tray. Nevertheless, cats require a certain level of social interaction with their owners and this requirement may be increased when their physical environment is restricted (see Chapter 4). Cats confined indoors spend proportionately more time with people than cats with access to the outdoors, which has been interpreted as cats seeking additional stimulation in an environment that is relatively less stimulating (Turner & Stammbach-Geering 1990). There are a number of websites that give advice to owners on how to enrich the indoor environment for their cat (for example www.nssvet.org/ici).

One reason for keeping cats indoors is to protect them from road traffic accidents. In a study of factors that may predispose cats to road traffic accidents in Cambridgeshire, United Kingdom, 115 owned cats that had been in a road traffic accident (RTA) were compared with a control population of 794 cats that had never been in a RTA (Rochlitz 2003a). RTA cats tended to be younger (46% were between 7 and 24 months of age), male (both neutered and entire, 62%) and non-pedigree (97%). For every one year increase in age, the odds of a road accident decreased by 16 per cent; the odds for males (entire and neutered) being in a road accident were 1.9 times the odds for females (entire and neutered), and the odds for pedigree cats were 0.29 those for non-pedigree cats. Proportionately more of the RTA cats lived in areas with higher levels of traffic and there was a trend for more accidents to happen during the night than the day (Rochlitz 2003b). In view of these findings, it may be appropriate to advise owners who are worried about their cat being involved in a RTA to adopt an older, neutered female, possibly of a pedigree breed, to keep it in at night and, if feasible, to live in an area with low levels of traffic.

The effects of predation by cats on wildlife should also be considered when evaluating indoor versus outdoor living. Solutions to enable the cat to benefit from outdoor access without risk to itself or others include restricting outdoor access to certain parts of the day, creating secure, yet stimulating and complex, cat-proof enclosures within a garden (for example, walk-in cages or modular structures extending through a garden, secure perimeter fences) or training a cat to go for walks on a leash, but they may not be possible in many situations. The effectiveness of bells or ultrasonic devices on collars, to alert wildlife to the cat's presence, appears to be limited.

With the current state of knowledge it is not possible to definitively say that confining cats indoors is preferable to allowing them outdoor access; each situation should be assessed individually, taking into account the cat, its owner and the local environment.

4.3.2 Requirements of Cats Kept in the Home

The home range of the pet cat confined indoors is inevitably very small compared to that of cats allowed to roam freely. Mertens and Schär (1988) recommend that an indoor-only cat should have access to at least two rooms. Bernstein and Strack (1996) described the use of space and patterns of interaction of 14 unrelated, neutered domestic cats, who lived together in a single-storey house at a density of one cat per 10 m^2, and did not have access to the outdoors. Most of the cats had favourite spots within the rooms that they used. Some individuals had their own unique place, but more commonly several cats chose the same favourite spot. These areas were shared either physically, by cats occupying the space together or, more often, temporally by cats occupying them at different times of the day. There was very little aggression and no fighting between the cats. Individuals seemed to peacefully co-exist with each other by avoiding each other for most of the time. Neutered males had an average home range of four to five rooms (out of 10), and neutered females a range of three to 3.6 rooms. While Mertens and Schär (1988) state that female cats may be more suited to an indoor existence than male cats, because feral males have bigger home ranges than feral females, it seems that both neutered males and neutered females can be successfully housed indoors providing there is sufficient quantity and quality of space and that they are used to these conditions from an early age.

Schroll (2002) suggests that there should be at least two types of resting places per cat, one on the floor enclosed by three sides and another elevated with a good view, and that this is particularly important to prevent behavioural problems in the multi-cat household. She also advises placing a scratching surface in more than one location, for example at places of entry and exit in the home, and also next to the resting or sleeping area, as these are locations that the cat will want to mark as part of its environment (see Chapter 4).

Litter trays should be positioned in a quiet place in the house, and cleaned at least once a day. Cats with easy access to the outdoors may not need a litter tray, although older cats, those who do not like to go out in bad weather, and cats that are unwell may require one.

Cats are often kept in the home together with another companion animal, such as a dog. Providing the cat is habituated to dogs, it is likely that the cat's social environment is enriched by this contact but there are few studies specifically examining the benefits to the cat from this interaction.

The more an owner responds to their cat the more likely it is to respond to them, and interactions initiated by the cat last longer than those initiated by the owner (Turner 1995a). In a study of interactions between cats and their owners in the home (Mertens 1991), cats in single-cat households stayed closer to their owners for longer, and had more social play and more interactions in general with them, than did cats in multi-cat households (see Chapter 3 for a discussion of the human-cat relationship).

Group-living cats lack distinct dominance hierarchies, signals for diffusing conflict and post-conflict mechanisms such as reconciliation (van den Bos & de Cock Buning 1994b; van den Bos 1998). They are not adapted to living in close proximity to each other, and in the wild would reduce the likelihood of aggression by dispersing or avoiding each other (Leyhausen 1979); this is often not possible in the multi-cat household. Recent evidence from clinical behavioural studies suggests that one of the major reasons for the development of behavioural problems, such as fearful or avoidance-related behaviours, are environmental stressors and that a high proportion of these relate to social factors (relationships with other cats or with humans) (Casey & Bradshaw 2000). The environment in which the cat is housed will affect the development and maintenance of these behavioural problems. Sibling pairs of cats have more amicable relationships than unrelated cats living together (Bradshaw & Hall 1999). The incidence of behavioural problems tends to increase when there are four or more cats in a household, particularly if the cats are unrelated. When cats are introduced to each other as adults, they may not regard each other as part of the same social group but they are forced to live together in relatively close proximity. When there is a high density of cats in a neighbourhood, cats may be frequently involved in aggressive encounters with others, and some cats may become frightened of going outdoors. Owners should be aware of the social dynamics between cats in their household, and be prepared to seek professional advice if it appears that the welfare of their cats is compromised (see Chapter 4).

5. CONCLUSIONS

Traditionally, because cats are small animals it was thought acceptable to house them in small enclosures. With the development of an understanding of the ethology of the cat, its evolutionary history and findings from studies of cats housed in different conditions, we recognise that this approach is unsatisfactory. A certain minimum amount of space, which is more generous than previously assumed, is needed in order to provide a good quality of space, enriched with places to hide and structures that enable use of the vertical dimension, a stimulating sensory environment and opportunities to explore and play. While most cats can live with conspecifics, in order to do so successfully they need enough space to be able to keep a certain distance, and to get out of sight of each other. The care and attention from the human caregiver is the crucial determinant of a cat's welfare in any housing condition.

While the aim in biomedical research should be to eventually end all live animal use, the way cats are kept in research facilities should be refined to optimise their welfare. This will also lead to better science, positively influence the public perception of research using animals and improve the cats' chances of being re-homed when the research is completed. For cats in shelters, providing an enriched environment that minimizes stress, reduces the risk of disease transmission, and increases the likelihood of being adopted, remains a formidable challenge. It is being increasingly recognised that some cats suffer from severe social stress when housed in groups, for example in multi-cat households and shelters. As more and more cats are kept as pets in many countries, studies of cats in the home setting and studies comparing the welfare of indoor-only cats with that of cats allowed outdoors, are urgently needed.

6. REFERENCES

Barry, K. J. and Crowell-Davis, S. L. (1999) Gender differences in the social behaviour of the neutered indoor-only domestic cat. *Applied Animal Behaviour Science* **64**, 193-211.
Benn, D. M. (1995) Innovations in research animal care. *J. American Veterinary Medical Association* **206**, 465-468.

Bernstein, P. L. and Strack, M. (1996) A game of cat and house: spatial patterns and behaviour of 14 cats (felis catus) in the home. *Anthrozoös* **9**, 25-39.

Bessant, C. (2002) *Boarding Cattery Manual.* Feline Advisory Bureau, Tisbury, Wiltshire, UK.

Bradshaw, J. W. S. (1992) *The behaviour of the domestic cat.* CAB International, Wallingford, Oxon.

Bradshaw, J. W. S. and Cameron-Beaumont, C. L. (2000) The signalling repertoire of the domestic cat and its undomesticated relatives. In Turner, D.C. and Bateson, P. (eds.). *The Domestic Cat: the biology of its behaviour*, 2nd edn., Cambridge University Press, Cambridge, pp. 68-93.

Bradshaw, J. W. S. and Hall, S. L. (1999) Affiliative behaviour of related and unrelated pairs of cats in catteries: a preliminary report. *Applied Animal Behaviour Science* **63**, 251-255.

Broom, D. M. and Johnson, K. G. (1993) *Stress and animal welfare.* Chapman and Hall Ltd., London.

Buffington, C. A. T. (2002) External and internal influences on disease risk in cats. *J. American Veterinary Medical Association* **220**, 994-1002.

Carlstead, K., Brown, J. L. and Strawn, W. (1993) Behavioural and physiological correlates of stress in laboratory cats. *Applied Animal Behaviour Science* **38**, 143-158.

Casey, R. and Bradshaw, J. W. S. (2000) Welfare implications of social stress in the domestic cat. Issues in Companion Animal Welfare, Amsterdam, International Society for Anthrozoology, p. 12.

CIEH Animal Boarding Establishments Working Party (1995) *Model licence conditions and guidance for cat boarding establishments (Animal Boarding Establishments Act 1963).* The Chartered Institute of Environmental Health, London.

Companion Animal Welfare Council (2004) *The report on companion animal welfare establishments: sanctuaries, shelters and re-homing centres,* Companion Animal Welfare Council, Devon.

Crouse, S. J., Atwill, E. R., Lagana, M. and Houpt, K. A. (1995) Soft surfaces: a factor in feline psychological well-being. *Contemporary Topics in Laboratory Animal Science* **34**, 94-97.

de Monte, M. and Le Pape, G. (1997) Behavioural effects of cage enrichment in single-caged adult cats. *Animal Welfare* **6**, 53-66.

DeLuca, A. M. and Kranda, K. C. (1992) Environmental enrichment in a large animal facility. *Laboratory Animals* **21**, 38-44.

Department for Environment, Food and Rural Affairs (2002) *The consultation on an animal welfare bill.* Animal Welfare Division, DEFRA, Page Street, London, pp. 1-7. www.defra.gov.uk/animalh/welfare/domestic/index.htm.

Durman, K. J. (1991) Behavioural indicators of stress. B.Sc. thesis, University of Southampton.

Evans, R. H. (2001) Feline animal shelter medicine. In August, J. R. (ed.). *Consultations in Feline Internal Medicine,* 4th Edn., W.B. Saunders Company, Philadelphia, pp. 571-576.

Gourkow, N. (2001) Factors affecting the welfare and adoption rate of cats in an animal shelter. M.Sc. thesis, University of British Columbia.

Hall, S. L. and Bradshaw, J. W. S. (1998) The influence of hunger on object play by adult domestic cats. *Applied Animal Behaviour Science* **58**, 143-150.

Hart, B. L. (1980) *Feline Behaviour: A Practitioner Monograph.* Veterinary Practice Publishing Company, Santa Barbara.

Haughie, A. (1998) *Cat Rescue Manual.* Feline Advisory Bureau, Tisbury, Wiltshire, UK.

Hawthorne, A. J., Loveridge, G. G. and Horrocks, L. J. (1995) The behaviour of domestic cats in response to a variety of surface-textures. In Holst, B. (ed.). *Proceedings of the second international conference on environmental enrichment,* Copenhagen Zoo, Copenhagen, pp. 84-94.

Holmes, R. J. (1993) Environmental enrichment for confined dogs and cats. In Holmes, R. J. (ed.). *Animal Behaviour-The TG Hungerford Refresher Course for Veterinarians, Proceedings 214,* Post Graduate Committee in Veterinary Science, Sydney, Australia, pp. 191-197.

Hoskins, C. M. (1995) The effects of positive handling on the behaviour of domestic cats in rescue centres. M.Sc. thesis, University of Edinburgh.

Hubrecht, R. C. and Turner, D. C. (1998) Companion animal welfare in private and institutional settings. In Turner, D. C. (ed.). *Companion Animals in Human Health,* Sage Publications Inc, Thousand Oaks, CA, pp. 267-289.

James, A. E. (1995) The laboratory cat. *ANZCCART News* **8**, 1-8.

Karsh, E. B. and Turner, D. C. (1988) The human-cat relationship. In Turner, D.C. and Bateson, P. (eds.). *The Domestic Cat: the biology of its behaviour,* 1st edn., Cambridge University Press, Cambridge, pp. 159-177.

Kessler, M. R. and Turner, D. C. (1997) Stress and adaptation of cats (*Felis Silvestris Catus*) housed singly, in pairs and in groups in boarding catteries. *Animal Welfare* **6**, 243-254.

Kessler, M. R. and Turner, D. C. (1999a) Socialisation and stress in cats (*Felis silvestris catus*) housed singly and in groups in animal shelters. *Animal Welfare* **8**, 15-26.

Kessler, M. R. and Turner, D. C. (1999b) Effects of density and cage size on stress in domestic cats (*Felis silvestris catus*) housed in animal shelters and boarding catteries. *Animal Welfare* **8**, 259-267.

Law, G., Boyle, H., Johnston, J. and Macdonald, A. (1991) Food presentation, part 2: cats. In *Environmental enrichment: advancing animal care.* Universities Federation for Animal Welfare, Potters Bar, Herts, pp. 103-105.

Leyhausen, P. (1979) *Cat behaviour: the predatory and social behaviour of domestic and wild cats.* Garland STPM Press, New York.

Lowe, S. E. and Bradshaw, J. W. (2001) Ontogeny of individuality in the domestic cat in the home environment. *Animal Behaviour* **61**, 231-237.

Markowitz, H. and LaForse, S. (1987) Artificial prey as behavioural enrichment devices for felines. *Applied Animal Behaviour Science* **18**, 31-43.

McCune, S. (1992) Temperament and the welfare of caged cats. Ph.D. thesis, University of Cambridge.

McCune, S. (1995) Enriching the environment of the laboratory cat - a review. In Holst, B. (ed.). *Proceedings of the second international conference on environmental enrichment,* Copenhagen Zoo, Copenhagen, pp. 103-117.

Mertens, C. (1991) Human-cat interactions in the human setting. *Anthrozoös* **4**, 214-231.

Mertens, C. and Schär, R. (1988) Practical aspects of research on cats. In Turner, D.C. and Bateson, P. (eds.). *The Domestic Cat: the biology of its behaviour,* 1st edn., Cambridge University Press, Cambridge, pp. 179-190.

Mertens, C. and Turner, D. C. (1988) Experimental analysis of human-cat interactions during first encounters. *Anthrozoös* **2**, 83-97.

National Research Council (1996) *Guide for the care and use of laboratory animals.* National Academy Press, Washington.

Newberry, R. C. (1995) Environmental enrichment: increasing the biological relevance of captive environments. *Applied Animal Behaviour Science* **44**, 229-243.

Ottway, D. S. and Hawkins, D. M. (2003) Cat housing in rescue shelters: a welfare comparison between communal and discrete-unit housing. *Animal Welfare* **12**, 173-189.

Patronek, G. J., Beck, A. M. and Glickman, L. T. (1997) Dynamics of dog and cat populations in a community. *J. American Veterinary Medical Association* **210**, 637-642.

Patronek, G. J. and Sperry, G. (2001) Quality of life in long-term confinement. In August, J. R. (ed.). *Consultations in Feline Internal Medicine,* 4th Edn., W.B. Saunders Company, Philadelphia, pp. 621-634.

Pedersen, N. C., Sato, R., Foley, J. E. and Poland, A. M. (2004) Common virus infections in cats, before and after being placed in shelters, with emphasis on feline enteric coronavirus. *J. Feline Medicine and Surgery* **6**, 83-88.

Podberscek, A. L., Blackshaw, J. K. and Beattie, A. W. (1991) The behaviour of laboratory colony cats and their reactions to a familiar and unfamiliar person. *Applied Animal Behaviour Science* **31**, 119-130.

Poole, T. B. (1997) Happy animals make good science. *Laboratory Animals* **31**, 116-124.

Randall, W. R., Cunningham, J. T. and Randall, S. (1990) Sounds from an animal colony entrain a circadian rhythm in the cat, *Felis catus* L. *J. Interdisciplinary Cycle Research* **21**, 55-64.

Reinhardt, V. and Reinhardt, A. (2001) Legal space requirement stipulations for animals in the laboratory: are they adequate? *J. Applied Animal Welfare Science* **4**, 143-149.

Rochlitz, I. (2003a) Study of factors that may predispose domestic cats to road traffic accidents: Part 1. *Veterinary Record* **153**, 549-553.

Rochlitz, I. (2003b) Study of factors that may predispose domestic cats to road traffic accidents: Part 2. *Veterinary Record* **153**, 585-588.

Rochlitz, I., Podberscek, A. L. and Broom, D. M. (1998) The welfare of cats in a quarantine cattery. *Veterinary Record* **142**,

Roy, D. (1992) Environmental enrichment for cats in rescue centres. B.Sc. thesis, University of Southampton.

Royal College of Veterinary Surgeons (2004) New practice standards scheme takes place. *RCVS news* **March**, 2.

Russell, W. M. and Birch, R. L. (1959) *The principles of humane experimental technique.* Methuen, London.

Schroll, S. (2002) Environmental enrichment for indoor cats as prevention and therapy- practical advice for quality of life. *Companion Animal Behaviour Study Therapy Group*, Birmingham, pp. 43-45.

Shepherdson, D. J., Carlstead, K., Mellen, J. D. and Seidensticker, J. (1993) The influence of food presentation on the behaviour of small cats in confined environments. *Zoo Biology* **12**, 203-216.

Siegford, J. M., Walshaw, S. O., Brunner, P. and Zanella, A. J. (2004) Validation of a temperament test for domestic cats. *Anthrozoös* **16**, 332-351.

Smith, D. F. E., Durman, K. J., Roy, D. B. and Bradshaw, J. W. S. (1994) Behavioural aspects of the welfare of rescued cats. *J. Feline Advisory Bureau* **31**, 25-28.

Turner, D. C. (1995a) The human-cat relationship. In Robinson, I. (ed.). *The Waltham book of human-animal interaction: benefits and responsibilities of pet ownership*, Elsevier Science Ltd., Oxford, pp. 87-97.

Turner, D. C. (1995b) The ethology of the domestic cat and its consequences for the human-cat relationship. *Proceedings of 7th International Conference on human-animal interactions: Animals, Health and Quality of Life*, p. 122.

Turner, D. C. and Stammbach-Geering, M. K. (1990) Owner assessment and the ethology of human-cat relationships. In Burger, I. H. (ed.). *Pets, benefits and practice*, British Veterinary Association Publications, London, pp. 25-30.

van den Bos, R. (1998) Post-conflict stress-response in confined group-living cats (*Felis silvestris catus*). *Applied Animal Behaviour Science* **59**, 323-330.

van den Bos, R. and de Cock Buning, T. (1994a) Social behaviour of domestic cats (*Felis lybica* f.*catus* L.): a study of dominance in a group of female laboratory cats. *Ethology* **98**, 14-37.

van den Bos, R. and de Cock Buning, T. (1994b) Social and non-social behaviour of domestic cats (*Felis catus* L.): a review of the literature and experimental findings. In Bunyan, J. (ed.). *Welfare and Science-proceedings of the fifth FELASA symposium,* Royal Society of Medicine Press Ltd., London, pp. 53-57.

Wells, D. and Egli, J. M. (2003) The influence of olfactory enrichment on the behaviour of captive black-footed cats, Felis Nigripes. *Applied Animal Behaviour Science* **85**, 107-119.

Wemelsfelder, F. (1991) Animal boredom : do animals miss being alert and active? In Rutter, S. M. (ed.). *Applied Animal Behaviour: past, present and future,* Universities Federation for Animal Welfare, Potters Bar, Herts, pp. 120-123.

Young, R. J. (2003) *Environmental enrichment for captive animals*. Blackwell Science Ltd., Oxford.

Chapter 8

DISEASE AND WELFARE

Kit Sturgess
Wey Referrals, 125-129 Chertsey Road, Woking, Surrey GU21 5BP

Abstract: Both infectious and non-infectious disease have a major impact on the welfare of cats. The likelihood of an individual developing a particular disease will depend on a variety of factors including age, exposure, genetic make-up, and general nutritional and health status. Disease can impact on an individual, and can also affect a group or a population of cats. With our current understanding, and with the multi-factorial nature of risk factors, preventing most non-infectious diseases is very difficult so we have to rely on early diagnosis and appropriate treatment. Infectious disease, however, may be more easily preventable using a variety of strategies including vaccination, reduction of exposure and improving resistance to infection. Despite the difficulties in recognising signs of pain, especially chronic pain, in cats, the prevention and treatment of pain is of major importance, and the development of effective easy-to administer analgesics, especially those for long-term use, should be a major research priority in veterinary medicine.

1. INTRODUCTION

1.1 Infectious and Non-infectious Disease

Disease has a significant impact on the welfare of cats and, ultimately, disease of one type or another will lead to their death. Infectious disease is particularly associated with young cats, whose immune system is not fully developed, and those kept in large groups. It tends to affect several individuals within a group and is often persistent within it; clinical signs, however, may be sporadic in their occurrence. The likelihood of an individual acquiring an infectious disease, and the severity of the disease, can be significantly altered by the management systems that are in place. Non-infectious disease, apart from a number of inherited conditions, tends to affect individuals and can be classified as degenerative (for example chronic renal failure, dental disease), inflammatory (inflammatory bowel disease), management-associated (due to poor diet or access to toxins), traumatic (for example a road accident), neoplastic (due to tumours or tumour-associated syndromes) and idiopathic (for example hyperthyroidism, an endocrine disease). Management can affect the prevalence and severity of non-infectious disease as well as the cost, success of treatment and outcome. However, many other factors affect the likelihood of an individual developing disease, in particular its genetic background.

Acute disease can cause profound suffering but, if self-limiting or appropriately treated, the duration of suffering will be short and there tends to be no, or minimal, long-term effects on the individual's welfare. However, the possibility that disease and the treatment of disease, such as hospitalisation and forced medication, may have long-term welfare implications cannot be excluded. Chronic disease may cause less obvious symptoms and be more difficult to recognise (particularly in cats as they are able to modify their lifestyle according to their reduced physical capability), but may have far more profound long-term effects on welfare.

1.2 Effects of Body Form and Breeding

In recent years, cat breeding has led to a significant variation from the standard body form of the domestic short hair cats (see Chapter 10). In many cases, the impact on welfare has been related to the degree of inbreeding and a relatively small genetic pool, leading to certain diseases becoming more prevalent in a particular breed. In some cases, the body form associated with a breed can cause long-term welfare problems, for example respiratory obstruction in brachycephalic cats such as Persians. The desire of some breeders to emphasise breed characteristics has led to 'ultra' type cats, where inherent body form-associated problems have been compounded. There has been a significant increase in breeding from cats that have severe mutations in order to create a unique breed, for example Munchkin, kangaroo or twisty cats. Such mutations invariably give rise to severe welfare issues, as they cause both chronic disease and prevent the cat from being able to express natural behaviours such as jumping.

1.3 Impact of Infectious Disease on the Individual, Group and Population

Disease can have an impact on the welfare of cats at a number of different levels. At the individual level there may be both the physical effects of disease and treatment, preventing a cat from following its natural behavioural patterns, and the psychological aspects of the disease and treatment. At the group level, disease can have an effect in three main ways: through spread to other individuals; altering the group dynamic so as to lead to disruption of social stability; and the whole group needing treatment and/or screening, whether or not all individuals are showing clinical evidence of disease. At the population level, disease has its effect mainly on the genetic make-up of that population. This can be a direct effect, associated with individuals dying from their disease or having passed on disease susceptibility to their offspring. An indirect effect can also occur, as attempts are made to breed away from a genetically associated disease with individuals being culled, or other, as yet unknown, disease problems being bred into the remaining group.

2. PRINCIPLES OF INFECTIOUS DISEASE PREVENTION

Disease prevention relies mainly on reducing risk factors. This can be achieved by a variety of methods such as providing correct nutrition (see Chapter 9), maintaining good general health, and in the case of infectious disease also by vaccination, screening and decreasing likelihood of exposure.

2.1 Vaccination

Vaccination, particularly primary vaccination, plays a central role in the control of infectious disease within populations. Not all individuals within a population, however, need to be vaccinated in order to achieve this benefit. Vaccination considerably reduces morbidity and mortality associated with an individual infectious agent, and therefore has a significant impact on improving welfare. Vaccination alone is not capable of controlling disease within a population, as the risk of an individual becoming infected by a particular agent is dependent upon the infectious dose, the virulence of the pathogen and the host immune response.

Vaccination serves to create an anamnestic response, such that when the vaccinated individual meets a field infection it produces a stronger and more rapid immune response than if unvaccinated (Schulz & Conklin 1998). If a protective immunity is achieved within the incubation period of the infection, the individual will show few, if any, clinical signs of disease. It is important to remember that vaccination does not prevent infection, hence vaccinated individuals can excrete the infectious agent, contaminating the environment, and they can also become carriers of the agent. However, in the majority of cases when a vaccinated individual becomes infected the duration and level of excretion of the infectious agent is less, and the environmental contamination is therefore reduced. Thus, indirectly, vaccination will reduce the infectious dose within the group. Vaccination may impact on virulence, as it tends to promote evolution of the infectious agent away from the vaccine strain(s) and this could potentially increase virulence. It is clinical disease, however, that directly affects welfare so vaccination is a positive benefit to the population and to the majority of individuals.

Vaccination is not without risks (Greene 1998), although these are generally to the individual and most commonly occur as vaccine reactions (see next section). There is also the potential for modified live vaccines to cause clinical disease, if the vaccine is administered incorrectly or given to an immunocompromised or pregnant individual. There are reports of certain breeds, and cats infected with feline immunodeficiency virus (FIV), being more susceptible to vaccine-induced disease (Buonavoglia *et al.* 1993). Under rare circumstances, vaccination can be a risk to the welfare of the population as a whole if a batch of vaccine becomes contaminated with another agent. For example, it has been suggested that the worldwide, simultaneous occurrence of parvovirus infection in dogs could have been the result of vaccine contamination.

2.1.1 Vaccine Reactions

Whether or not to vaccinate is a balance between risks and benefits; risks to the individual of vaccination are primarily those of a vaccine reaction. Should large number of individuals within a population remain unvaccinated then the potential exists for an epidemic to occur.

Vaccine reactions are the most common adverse drug reaction reported to the Veterinary Medicines Directorate in the United Kingdom. To some extent vaccine reactions are to be expected, as some sort of response by the individual is necessary in order to stimulate the immune system or the vaccine is unlikely to be efficacious. What is classified as a vaccine reaction is also unclear, as low-grade malaise of less than 24 hours duration is not uncommon but many would not regard this as a vaccine reaction. A crude estimate is that around 3% of vaccinations result in 'significant' vaccine reactions, i.e. a reaction beyond 24 hours of low-grade malaise. The impact on the welfare of the individual, in the majority of cases, is short-lived. Recently, however, vaccination in cats has been linked to the development of vaccine-associated sarcomas (invasive soft tissue tumours) that develop at the site of vaccination, usually in the neck region of the cat. The risk has been calculated to be 1 in 10,000 (0.01%) vaccinations (Hendrick 1999). The consequences of a vaccine-associated sarcoma are severe and potentially fatal. This has prompted the American Association of Feline Practitioners to recommend that vaccines should be given in the distal limbs of the cat (Levy *et al.* 2001), as a limb can be amputated should a sarcoma occur. Further, different vaccines are given at different limb sites in order to try and identify those vaccines associated with the development of sarcomas.

2.1.2 Infectious Diseases against which Vaccination is Available

There are an ever-increasing number of vaccines available on the veterinary market for use in cats. These include vaccines against the following infectious agents: feline parvovirus (also known as feline infectious enteritis, feline panleukopenia); feline herpes virus-1 (feline rhinotracheitis); feline calicivirus; rabies virus; *Chlamydophila felis* (Chlamydiosis, *Chlamydia psittaci*); feline leukaemia virus (FeLV); *Bordetella bronchiseptica*; feline coronavirus (feline infectious peritonitis); *Microsporum canis* (Ringworm, this is a post-exposure vaccine); *Borrelia burgdorferi* (Lyme disease); feline immunodeficiency virus (FIV) and *Toxoplasma gondii* (toxoplasmosis). Vaccines against all these agents are available in the United States, but in the United Kingdom the last five are not.

Few individuals are going to be vaccinated using all potential vaccines, hence it is necessary to decide which vaccines are most suitable for each individual. Likelihood of exposure, severity of disease, known risks of the particular vaccine and previous response of the individual to vaccination should be considered. There are also the effects of a particular infectious agent on the cat population as a whole to be taken into account. Whilst it would be ideal to risk assess each individual, this is rarely possible because so many factors are unknown. This has lead to the concept of 'core' and 'non-core' vaccines. In the United Kingdom, feline parvovirus and the respiratory viruses (feline rhinotracheitis and feline calicivirus) and sometimes feline leukaemia virus are considered core vaccines, while Chlamydophila and Bordetella are non-core and rabies virus vaccine is given according to need, for example if the cat is travelling abroad.

Concomitant with the debate on which vaccines to use, has been the debate on the frequency with which boosters should be given (particularly in view of the risk of vaccine-associated sarcoma). There is little evidence available on the true duration of immunity engendered by different vaccines. All vaccines will protect the majority of individuals for a minimum of one year, although it has been suggested that booster vaccinations against respiratory viruses should be given six-monthly in cats at high risk of infection. It is also clear that the duration of immunity following some vaccines, for example parvovirus, is considerably longer than one year depending on the vaccine used and the degree of exposure in the field. This has led to the recommendation that boosters should be given every three years to cats (Levy *et al.* 2001). However, clear scientific evidence in support of this approach is lacking. Again, an individual risk assessment, as far as is possible, is the most suitable approach when deciding how often to administer boosters.

2.2 Screening for Infectious Disease

Cats may be screened for infectious disease in a number of situations, particularly when cats are kept in groups. For example, a breeding cattery may be free of a particular disease so incoming cats should not have that disease if allowed to enter the cattery. The veterinary history of a cat adopted from a shelter may not be known, but the adopter may want to ascertain that the new cat will not pose a risk to existing cats in the house.

There are few reasons to screen for bacterial disease in cats, unless a particular group is to be kept disease-free, for example a breeding cattery with a *Bordetella bronchiseptica* negative status. Faecal screening for some pathogenic bacteria or protozoa may be indicated in specific instances. With the recent availability of a polymerase chain reaction (PCR) test for haemobartonellosis (a mycoplasma agent that is the cause of the disease feline infectious anaemia) (Tasker *et al.* 2003), screening for this infection may become more widespread.

In cats, screening for viral infections is most commonly performed.

2.2.1 Retroviruses

Screening for retroviruses (FIV and FeLV)) in healthy individuals is fraught with difficulties, both in terms of the sensitivity and specificity of the tests available and the action to be taken should an individual be positive (Jarrett *et al.* 1991; Robinson *et al.* 1998). Nevertheless, screening for retroviruses is widespread in veterinary medicine. Depending on the prevalence of infection, enzyme-linked immunosorbent assay (ELISA) or immunochromatography testing in healthy individuals will have a false positive rate of between 30 and 50%. Therefore, a cat should not be euthanased on the basis of a single positive test. Sensitivity can be improved by testing 'high-risk' groups where the prevalence of infection should be higher. Traditionally, these groups were thought to be rescue and feral cats; however, a higher prevalence of infection in feral cat populations has not been documented in a number of studies. In many cases, the prevalence of infection appears to be a local phenomenon. Higher false positive rates have also been documented in cats that have been recently vaccinated. Some tests will also give a positive result if the cat has anti-mouse antibodies in its blood or is infected with spumavirus (a related retrovirus). A variety of other methods such as PCR, virus isolation, western blotting or immunofluoresence testing can be used to detect FeLV (Herring *et al.* 2001) and can be used to confirm a positive ELISA result; alternatively, the screening test can be repeated 6 to 8 weeks later. Discordant results will occur and should be investigated further. In the majority of cases, results will eventually become concordant but a number of individuals will remain persistently discordant and their true virus status unknown.

Cats positive for FIV on screening will often have a long asymptomatic period (usually 2 to 5 years but longer periods are reported) before they show signs of illness, and the risk of virus transmission to other individuals in a group is thought to be low. The main value in knowing the FIV status is for the individual's benefit, allowing prompt and aggressive treatment of other infectious disease that may arise in a cat that is immunocompromised due to FIV infection. Responsible management would also include keeping the FIV positive cat indoors as far as is practicable, to reduce spread of infection through fighting and biting. Entire adults should be neutered, as this reduces fighting in male cats and the risk of transplacental spread in females. Approximately 25 to 30% of kittens born to an infected female are likely to be FIV positive (O'Neill *et al.* 1995), but infection rates will depend on the stage of infection of the queen and the FIV strain involved.

Cats positive for FeLV on screening pose a greater risk to FeLV negative individuals, as the virus is spread by social contact more easily. Whilst vaccination against FeLV may help protect FeLV negative cats from infection, the preventable fraction is significantly less than 100% (Sparkes 2003) (the preventable fraction is the percentage of cats that would be expected to become infected that do not, following vaccination). The prognosis for FeLV positive cats is also more guarded, with more than 80% likely to die within three years of their positive FeLV status being detected. Options for FeLV positive cats include creating a FeLV positive group of cats within a home, keeping infected cats as single cats in a household and euthanasia.

2.2.2 Other Viruses

In one survey, where samples to screen for respiratory viruses were taken from apparently healthy cats at a cat show, approximately 30% of cats less than one year of age were positive for feline calicivirus (FCV). By comparison, 1% of the cats were positive for feline herpes virus (FHV-1) (Coutts *et al.* 1994). Even though the prevalence of FCV and FHV-1 are thought to be similar, screening for FHV-1 is insensitive because the virus is sequestered in the trigeminal ganglion in carrier cats and is excreted only intermittently, while FCV tends to be excreted continuously (see section 3.3). With FCV infection, however, there are some cats that shed low numbers of virus and need to be sampled on more than one occasion to accurately demonstrate their status. Under most circumstances there is little justification for screening for FCV, given that a large number of cats will be positive. Further, many strains are of low pathogenicity and therefore constitute a minor risk to the cat. Healthy cats should not be euthanased if they are positive for FCV or FHV-1.

The existence of carriers of feline parvovirus (FPV) has recently been suggested. Such individuals could potentially represent a risk to unvaccinated cats within the group that are immunologically naive. The sensitivity of faecal examination for identifying carriers of the virus is unknown. The risk of parvovirus infection is probably better reduced by quarantine (for those individuals incubating primary disease) and vaccination, than by screening for carriers.

2.3 Reduction of Exposure

Disease can be caused by exposure to pathogens and to potentially harmful toxins. Ways to reduce the likelihood of exposure of cats to pathogens are listed in Table 1, and ways to reduce the likelihood of exposure to toxins are listed in Table 2.

For the pet cat, practical solutions to reduce exposure depend on decisions made regarding whether the cat is allowed access beyond the house and garden. Cats may be kept wholly indoors, provided with outside pens, the garden may have a perimeter fence that is cat-proof both from ingress and egress, or the cat may be allowed to roam freely. For cats kept in large groups, such as in catteries and shelters, the control of infectious disease is a major challenge; this is discussed further in section 4.4.

Table 1. Ways of reducing the exposure of cats to pathogens

Decreased contact with other cats, in particular those that are likely to be carriers, incubating disease, or are overtly affected
Spacial or appropriate chronological separation from areas where potentially infectious individuals have been
Avoiding areas likely to be contaminated, for example catteries, rescue centres, veterinary surgeons' waiting rooms
Disinfection of the environment
Vaccination of likely contacts
Quarantine of individuals likely to be infectious
Management practices that reduce the likelihood of spread on inanimate objects (such as clothes, food bowls, grooming and cleaning equipment)

Table 2. Ways of reducing the exposure of cats to toxins

Careful storage and disposal of potential toxic substances
Reduced access to areas (for example neighbours' gardens) where control of potential toxins is unknown
Rapid removal of any potential toxins from the cat's coat to prevent ingestion by grooming
Careful selection and storage of food substances to prevent contamination
Reduced access to food substances that can not be controlled, for example food left out by others, dead prey, live prey species potentially containing toxins

3. METHODS OF INFECTIOUS DISEASE SPREAD

Understanding how infectious disease is spread is key to understanding how to control it, particularly when cats are kept in groups. There are two major ways that infection is spread: horizontal transmission between cats and vertical transmission between a queen and her kittens *in utero*. A carrier of an infectious disease is an animal that does not show clinical signs of the disease but whose body harbours the disease-producing organism and may continue to excrete it. Carrier animals are of great epidemiological importance in the spread of infectious disease in cats.

3.1 Horizontal Transmission

Infection can be transmitted both in the acute phase, when the cat is obviously unwell, and during the incubation period before it has become ill. Recovered cats can become carriers, remaining healthy but continuing to spread infection to susceptible individuals. Horizontal transmission can be by direct cat-to-cat contact or via inanimate objects (indirect transmission).

When disease is spread by direct contact, a part of the body of one animal meets a body part of another animal, for example when skin surfaces come into contact, when one animal licks or grooms another, or during fighting. (Venereal transmission of disease, involving direct contact between the reproductive organs, occurs in dogs and cats but is not a significant route of infection in the United Kingdom). Infectious agents that are spread by direct contact are frequently fragile organisms; they are easily killed by heat, light, desiccation and disinfectants. Disinfection is not, however, a major method of control in such infections. Another method of spread by direct contact is airborne transmission, where the infection is spread in droplets produced during coughing or sneezing. Airborne transmission is particularly important in the spread of respiratory diseases.

When disease is spread by indirect contact, two or more animals come into contact with the same inanimate object, or fomite, such as bedding material or feeding bowls. Pathogenic organisms are spread via this inanimate object. Some organisms can remain viable in the environment for long periods of time, particularly in dark, damp conditions and where the object has been contaminated with faecal or other organic material. Usually contact with the inanimate object occurs within a short time after contamination; however some infectious agents, such as feline parvovirus, can survive for very long periods in the environment. Infectious agents that rely on indirect spread are generally hardy and more difficult to kill with disinfectants.

Some infectious agents do not pass directly from one individual to another but spend part of their lifecycle on or in another host requiring a vector for transmission, for example the tapeworm *Dipylidium caninum*, which affects cats, uses small rodents as an intermediate host and fleas as the vector.

3.2 Vertical Transmission

Feline parvovirus can be spread vertically if the queen becomes infected whilst she is pregnant. The outcome of such an infection will depend on the stage of the pregnancy. In the case of feline parvovirus, infection can cause a variety of problems including abortion, the birth of mummified kittens, and underdevelopment of the cerebellum, where the kittens are born alive but are poorly co-ordinated due to the cerebellar hypoplasia. Feline leukaemia virus and FIV can also be transmitted vertically.

3.3 Carrier Cats

A carrier cat can be placed in one of four categories, depending on whether it has shown clinical evidence of disease (convalescent or healthy) and on its level of excretion of infectious agent (continuous or intermittent), according to the definitions below:

- Convalescent - individuals who have had the disease, with the usual clinical signs, but who do not rid themselves of the organism completely for a long time; in some cases the organism persists in the animal for the rest of its life.
- Healthy - individuals who have never shown typical clinical signs of the disease. They possess an innate immunity to the organism which is sufficient to prevent clinical signs but not sufficient to prevent infection. Vaccinated animals can become carriers in this way. Healthy carriers can excrete the organism continuously or intermittently without becoming clinically affected themselves.

- Continuous excretors - individuals who continuously excrete the infectious agent and can infect other animals at any time. They are easier to identify than intermittent excretors.
- Intermittent excretors - individuals who only excrete organisms under certain circumstances, usually following periods of stress such as parturition, lactation, rehoming, or the use of immunosuppressive drugs.

Following infection with respiratory viruses, approximately 80% of cats with FHV-1 are thought to become intermittent carriers and are carriers for the rest of their life. Following FCV infection, 50% of cats become continuous excretors and are still excreting virus 90 days post-infection; however, they usually stop excreting after a period of time.

4. METHODS OF INFECTIOUS DISEASE CONTROL

Approaches to disease control include attention to hygiene, reduction of stress factors that may exacerbate disease, isolation and quarantine of potentially infected cats or cats with an unknown vaccination or health history, and measures specific to a particular group of cats or situation.

4.1 Hygiene

Hygiene plays a crucial role in the control of diseases, in particular those spread by indirect contact. In order for hygiene measures to be effective it is essential that all personnel adhere to the disinfection protocols, that the disinfectants used are appropriate ones for the infectious agent and are used at the correct concentration and in the correct manner, and that the disinfectant is safe to use in the environment where cats live.

Hygiene should encompass the cleaning of the living space of the cat, fomites (bowls, litter trays, grooming equipment) and personnel as they move from cat group to cat group. A number of disinfectants, in particular those containing phenolic compounds, are toxic to cats (Liao & Oehme 1980).

4.2 Reduction of Stressors

Stress can increase the likelihood of an individual developing clinical signs of infectious disease due to effects on the host immune response. It can also affect the severity and duration of the clinical signs. Short-term stress results in increases in the hormone cortisol that do not have a significant effect on the host immune response, and may even cause some enhancement. Long-term stress, however, tends to reduce resistance to disease by compromising the immune system due to the chronic release of hormones (such as cortisol) and cytokines (see Chapter 2). Stress can also be an important factor in the development of non-infectious diseases such as idiopathic feline lower urinary tract disease (Cameron *et al.* 2004), and of behavioural problems such as inappropriate elimination (urination or defecation) (see Chapter 4). Stress can also exacerbate infectious disease indirectly, through the development of stress-induced non-infectious disease or through poor nutritional intake.

4.3 Quarantine and Isolation

Quarantine and isolation are effective methods of reducing exposure to infectious disease. Quarantine is used before introducing a new individual to a group, as it allows time for infectious diseases that the individual is incubating to become clinically apparent. Whilst a quarantine period of 10 to 14 days is suitable for the control of many infectious diseases, it is not sufficient for diseases with a prolonged incubation, particularly rabies, FeLV or FIV. Further, it will not identify asymptomatic carriers. Quarantine is important especially when new, young cats are continuously being added to the group and where the background of the cat is unknown. In many instances, however, it is the new individual that is at higher risk of becoming infected from the group than vice versa. Quarantine also allows time for vaccination to become effective. Following quarantine, an individual can be exposed to potential infections in a controlled manner, in the hope that immunity occurs with the minimum of clinical signs. Welfare aspects should always be borne in mind when an individual is placed in isolation or quarantine.

Isolation of subsets of cats within a group can be of value if:

- Disease has occurred in one part of the premises but as yet not all cats have been exposed.
- Individuals have different disease status e.g. isolating FeLV positive from FeLV negative cats.
- Queens need to be separated from other cats from the time that they are due to kitten until after the kittens have been vaccinated.
- Kittens need to be separated from a queen likely to be a carrier, in the period between waning of their passive immunity and vaccination.

4.4 Disease Prevention and Control in High Risk Groups

While basic methods of disease prevention and control are applicable to all situations, additional approaches will vary depending on the way groups of cats are kept. They mainly rely on identifying potentially infectious cats and then either preventing them from entering the group or placing them in isolation and quarantine. Existing members of the group can be protected by vaccination and ensuring their immune system is effective, through good nutrition and the reduction of intercurrent disease and stress.

4.4.1   Multi-cat Households

For cats kept as pets in multi-cat households, the disease risk is relatively low if the group is stable and there is sufficient room for the number of cats. The risks can be further reduced by knowing the infectious diseases that exist within the group, as this allows risk benefit decisions to be made for an individual cat and for the group as a whole. It also allows the risk of introducing a new individual to the group to be assessed in terms of the likelihood and consequences of new infection being introduced and the risks to the new individual. Screening and immunization, together with quarantine and/or isolation of the new arrival, are most appropriate.

4.4.2   Breeding and Boarding Catteries

In breeding and boarding catteries, there is a major potential risk for infectious disease as cats are continually entering the premises, having arrived from environments over which the owner of the cattery has little control.

4.4.2.1 Breeding catteries
The major risk of infection in a breeding cattery is from visiting queens, new acquisitions and the continual or intermittent presence of immunologically naive individuals (kittens) within the group. Isolation of the various groups (particularly kittens) together with immunization, screening and quarantine of new members is required. This should be combined with knowledge of the disease state of the group as a whole. In most breeding establishments cats are kept indoors or in outdoor pens, so the risk of infection being introduced from cats outside the group is low. Nevertheless, periodic screening is advisable and vaccination to increase levels of immunity within the group is important. Not uncommonly there is a pet cat within a breeding cattery, and in the author's experience this cat can often be the source of infection as it is the only cat allowed outside and it may well not have been screened prior to introduction to the group.

4.4.2.2 Boarding catteries

The disease status of individuals entering a boarding cattery is generally unknown (even though the cat is usually vaccinated) and there is a continual movement of new individuals in and out of the cattery. Disease control has to rely on hygienic management practices and the construction of the premises to minimise disease spread, for example the placing of sneeze barriers between pens. Vaccination will increase resistance but does not prevent the cat being a carrier of infectious disease. The stress of boarding may well cause healthy carriers to become active excretors (see section 3.3). It is vital that cats are not moved from cage to cage to facilitate cleaning. Ideally each pen should have dedicated cleaning equipment, litter trays, food bowls and other items that are sterilised or discarded after the individual(s) have left.

4.4.3   Cat Shelters

Infectious disease is a major problem within many cat shelters (Cave *et al.* 2002) (Figure 1). Surveys of infection rates in individuals have shown a 2 to 3 fold increase in infection rates in cats that have entered rescue shelters compared to their status on admission (Pedersen *et al.* 2004). In some instances, it can be argued that the welfare of cats brought into a shelter with a significant infectious disease problem may be worse than that of cats left to fend for themselves, for example in feral cat populations.

It is virtually impossible to prevent the entry of infectious disease in rescue facilities and shelters; hence it is essential that buildings and management practices are designed so as to limit the spread of infectious disease. Practically this means:

- Quarantine for new arrivals.
- Maintaining cats in small, stable groups that are allowed to dwindle as cats are rehomed. Small groups should not be combined for easier management.
- Housing kittens together and away from adult cats.
- Vaccination, where financially practical, should be given 7 to 10 days after arriving when general health and disease status have been evaluated.
- Long stay cats should be housed separately from short stay cats
- Particular care should be taken to restrict access of any 'shelter cats' as they are sometimes allowed to roam free and may carry infection into or spread infection around the rescue facility.
- Accommodation should be designed to allow easy cleaning and prevent spread of disease to other cages (sneeze barriers, wide corridors, anteroom for cleaning and grooming equipment).

- Equipment should be specific to each individual or group and hygiene measures such as boot dips adopted.
- Cats should be cleaned in the order from the least likely to be infectious to the most likely group.
- Staff should be fully conversant with hygiene practices.

Environmental enrichment and other husbandry techniques to reduce stress should be practised (see Chapter 7).

Figure 1. Infectious disease due to respiratory viruses is a major problem in many shelters, and young animals are most susceptible.

4.4.4 Feral Populations

Surprisingly, stable feral populations are often remarkably free of many infectious diseases. Disease control can be achieved by a trapping, neutering and returning policy together with testing for FeLV and FIV at the time of neutering (see Chapter 6).

5. NON-INFECTIOUS DISEASE

The majority of sick pet cats presenting to veterinary surgeons are suffering from non-infectious disease. This is different from the situation in rescue facilities and some feral cat populations where the incidence of infectious diseases is likely to be greater. Of non-infectious disease, the most commonly reported in surveys of cats attending veterinary surgeries are dental disease, trauma, chronic renal failure (CRF) and gastrointestinal disease. Other common conditions requiring veterinary attention include feline lower urinary tract disease, hyperthyroidism and neoplasia (tumours or tumour-associated diseases). Dental disease, CRF, hyperthyroidism and neoplasia are primarily diseases of older cats. Trauma and gastrointestinal disease may affect cats of any age, although road traffic accidents affect mainly young cats (Rochlitz 2003). Feline lower urinary tract disease is more commonly reported in young to middle-aged cats. Screening will allow earlier detection of degenerative diseases. However, apart from dental disease, there is little information on the benefit of interventional therapies in delaying the onset of clinical disease. Notwithstanding this, preventative health care is important and screening for disease should be encouraged.

5.1 Screening for Non-infectious Disease

The biggest challenge with screening in any population is encouraging presentation of the cat to the clinician to allow screening to be conducted. Many practices offer annual health checks that are usually combined with vaccination, so that no specific charge for the health check is made. Unfortunately, relatively few cats are presented for annual booster vaccination, particularly as they get older. The other opportunity for health screening is when the cat is brought to the clinic for a specific reason, thereby allowing discussion of more general health issues.

Screening can be performed at a number of levels, and most commonly involves history taking, physical examination and blood tests. Physical examination, as a method of health screening, is a standard assessment of an individual but is relatively insensitive in its ability to diagnose occult disease. This has led to the use of other methods to minimise risk and detect disease as early as possible, in particular blood tests (for example prior to anaesthesia). At what age and how frequently an individual should be screened in order to deliver maximum health benefits is unknown. Many screening tests are relatively insensitive; for example, over 75% of renal mass is lost before blood concentrations of urea and creatinine (metabolites excreted by the kidneys) begin to rise. Further, little work has been performed to demonstrate which intervention, and at what stage, would benefit the individual. Before undertaking screening tests, a clear plan for the interpretation of the results and the action to be taken, if results are abnormal, should be established.

5.2 History Taking and Physical Examination

The skill of history taking and physical examination is one that all veterinary health professionals need to develop. Depending on the experience and training of the professional, history taking and physical examination may lead to a diagnosis. If not, it will significantly narrow the field of likely conditions to be considered and help direct further investigation. Many owners are highly observant and pick up very subtle changes that would not be apparent to the veterinary professional that does not have intimate knowledge of the individual cat; these observations should not be disregarded. The health professional's role is to prompt information from the owner by asking questions in a structured way, and to interpret the observations that have been made. It is often surprising that many owners have noticed overt clinical signs in their cats but have not pursued them further.

During history taking, key questions include those about:

- Appetite – change, duration and attitude towards food, for example if the appetite is decreased, is the cat asking for food and then not eating normally or is it less interested in food.
- Weight – visual changes of weight can often be missed, particularly in longhaired cats, but owners will often notice the change when they pick the cat up.
- Activity – if asked as a direct question, changes in activity are often not mentioned by owners. However, owners will often have noticed whether the cat is in the house more or sleeping more, and whether the distance over which it appears to roam is reduced. How well the cat is jumping can also be revealing.
- Behaviour – is the cat doing the same things it used to do? Has there been a change in the amount of attention seeking?

It is vitally important to ask about appetite, activity and behaviour as some of the changes that the owner may see as desirable or positive, such as an increased appetite, becoming more homely or more affectionate, may indicate problems such as hyperthyroidism or cardiovascular disease. In older cats, it can sometimes be very hard to distinguish changes that are associated with the normal aging process from changes indicative of developing disease. In general, aging changes are slowly progressive with no clear start point and should be at a level that is within the expected boundaries for a cat of that age.

Following the history, a thorough physical examination should be performed and recorded. A minimum recorded database should include temperature, pulse and heart rate, respiratory rate, colour of mucous membranes, capillary refill time, oral health and body weight. Repeat examinations identifying trends are a much more sensitive way of detecting low grade disease than an examination at a single time point, when the reference is whether an individual falls within the normal population range. Thorough examination of older cats may well reveal abnormalities that then need to be interpreted according to previous findings and the clinician's experience. Subclinical conditions may be associated with non-specific or normal historical and physical findings. In these circumstances, screening blood and urine tests may be of value.

6. THE RECOGNITION AND TREATMENT OF PAIN

Pain is an important welfare issue in all species, not least cats. However, effective pain relief can only be achieved and maintained when the signs of pain are recognised. Recognising signs of pain is complicated by the sedative action of many analgesics in veterinary use.

It is usually relatively easy to assess a cat's response to acute pain. Cats undergoing minor trauma respond by flinching, vocalization, attempts to escape or, occasionally, aggression. More severe injury usually results in the cat hissing, spitting, becoming aggressive or making vigorous attempts to escape. Following the acute response to major trauma, signs that the cat continues to be in pain become less obvious. Typically, the cat will become withdrawn and immobile; vocalization is rare but the cat will appear tense and distant and may emit occasional low growls. There is significant variation between cats and some will continue to spit and hiss whenever they are approached. A rapid respiratory rate is not uncommon and appears to be a pain response, as respiratory rate will frequently fall following analgesia. Later in the time course following acute injury, most cats will attempt to hide and show a marked reduction in appetite.

Following acute trauma the existence of pain is rarely in doubt, unless the cat is not found until some time after the event and external evidence is no longer apparent. The major clinical decision is not whether to give pain relief but what type of pain relief is most appropriate, and judging how long pain relief is necessary. In the majority of cases, continuing pain relief until near normal behaviour returns is appropriate. Administration of analgesics can be a challenge in cats, and novel routes have been investigated such as giving opioids (buprenorphine) intra-orally (Robertson *et al.* 2003).

Chronic pain is much more difficult to recognise in cats. As a species they are generally secretive about any form of incapacity, and will attempt to hide the fact that they are not 100% fit by altering their behaviour. Typical signs of chronic pain in cats include reduced activity, hiding, decreased interest and response to surroundings and weight loss due to inappetence (these signs are similar to those of chronic stress; see Chapters 2 and 4). Unfortunately, such signs are non-specific and can be associated with other disease processes where pain is not thought to be a significant feature. Pain can be difficult to localise, either because the cat fails to react when the focus is palpated or because the cat reacts wherever it is touched or handled. Localised pain may be seen as abnormalities of posture or prehension, lameness or stiffness or reluctance to perform a specific activity such as jumping. Chronic long-term pain, such as that caused by degenerative joint disease, is likely to have a more significant impact on the welfare of cats than is currently recognised.

Historically, analgesia has been underused in cats except following major orthopaedic procedures. This attitude is changing, and has been associated with a better understanding of pain management in cats and with an increase in the number of women in the profession (Dohoo & Dohoo 1996). Because of their unique metabolism and poor ability to glucuronidate drugs, non-steroidal anti-inflammatory drugs (NSAIDs) have been avoided in cats due to their perceived toxicity. The use of opioids was considered inappropriate, as hyperexcitability is common in cats given high doses, and individual cats can become excited at relatively low doses too. These views have now changed and compounds are in widespread use (albeit with precise dosing protocols) providing effective, short-term pain relief (Taylor *et al.* 2001).

In the United Kingdom, a number of analgesics are licensed for short-term use in cats and are either opiate-based or NSAIDs. These drugs have mainly been evaluated in studies involving post-operative pain relief (Balmer *et al.* 1998; Slingsby & Waterman-Pearson 2000; Lamont 2002). The use of other drugs which have analgesic activity, such as ketamine and medetomidine, is appropriate in some cases, and the value of local anaesthesia should not be overlooked. In extreme cases, euthanasia should be considered as a method of relieving intractable pain.

Analgesia for chronic pain is usually provided by NSAIDs, although they are not licensed for long-term use in cats in the United Kingdom. If NSAIDs are insufficient or inappropriate, opioids such as fentanyl patches can be used (Egger *et al.* 2003).

7. CONCLUSIONS

Infectious and non-infectious diseases can have major impacts on the welfare of cats at the level of the individual, group and population. Recent developments in vaccinology and vaccine protocols should serve to reduce the incidence of clinical disease, though infectious disease will remain a difficult problem in situations where cats are kept in large groups and the composition of these groups is unstable. Improvements in health care, screening and treatment will also reduce the effects of non-infectious disease on welfare by decreasing morbidity and mortality. However, there remain a large percentage of cats that do not have ready access to veterinary care; in these cats the effects of disease, especially infectious disease, can be severe. Studies on behaviours associated with pain in cats are needed, in order to develop better methods of identifying acute (Dixon *et al.* 2002) and chronic cases, and of assessing the effects of analgesics. With the increasing popularity of cats as companion animals in many countries, and their increased longevity associated with improved health care, the development of effective, easy-to-administer analgesics that are safe for long-term administration is urgently required.

8. REFERENCES

Balmer, T.V., Irvine, D., Jones, R.S., Roberts, M.J., Slingsby, L., Taylor, P.M., Waterman, A.E. and Waters, C. (1998) Comparison of carprofen and pethidine as postoperative analgesics in the cat. *J. Small Animal Practice* **39**, 158-164.

Buonavoglia, C., Marsilio, F., Tempesta, M., Buonavoglia, D., Tiscar, P.G., Cavalli, A. and Compagnucci, M. (1993) Use of a feline panleukopenia modified live virus vaccine in cats in the primary-stage of feline immunodeficiency virus infection. *Zentralblatt fur Veterinarmedizin, Reihe B.* **40**, 343-346.

Cameron, M.E., Casey, R.A., Bradshaw, J.W., Waran, N.K. and Gunn-Moore, D.A. (2004) A study of environmental and behavioural factors that may be associated with feline idiopathic cystitis. *J. Small Animal Practice* **45**, 144-147.

Cave, T.A., Thompson, H., Reid, S.W., Hodgson, D.R. and Addie, D.D. (2002) Kitten mortality in the United Kingdom: a retrospective analysis of 274 histopathological examinations (1986 to 2000). *Veterinary Record* **151**, 497-501.

Coutts, A.J., Dawson, S., Willoughby, K. and Gaskell, R.M. (1994) Isolation of feline respiratory viruses from clinically healthy cats at UK cat shows. *Veterinary Record* **135**, 555-556.

Dixon, M.J., Robertson, S.A. and Taylor, P.M. (2002) A thermal threshold testing device for evaluation of analgesics in cats. *Research in Veterinary Science* **72**, 205-210.

Dohoo, S.E. and Dohoo, I.R. (1996) Postoperative use of analgesics in dogs and cats by Canadian veterinarians. *Canadian Veterinary J.* **37**, 546-551.

Egger, C.M., Glerum, L.E., Allen, S.W. and Haag, M. (2003) Plasma fentanyl concentrations in awake cats and cats undergoing anesthesia and ovariohysterectomy using transdermal administration. *Veterinary Anaesthesia and Analgesia* **30**, 229-236.

Greene, C.G. (1998) *Infectious diseases of the dog and cat*. W.B. Saunders, Philadelphia, pp. 737-744.

Hendrick, M.J. (1999) Feline vaccine-associated sarcomas. *Cancer Investigations* **17**, 273-277.

Herring, I.P., Troy, G.C., Toth, T.E., Champagne, E.S., Pickett, J.P. and Haines, D.M. (2001) Feline leukemia virus detection in corneal tissues of cats by polymerase chain reaction and immunohistochemistry. *Veterinary Ophthalmology* **4**, 119-126.

Jarrett, O., Pacitti, A.M., Hosie, M.J. and Reid, G. (1991) Comparison of diagnostic methods for feline leukemia virus and feline immunodeficiency virus. *J. American Veterinary Medical Association* **199**, 1362-1364.

Lamont, L.A. (2002) Feline perioperative pain management. *Veterinary Clinics of North America: Small Animal Practice* **32**, 747-763.

Levy, J., Richards, J., Edwards, D., Elston, T., Hartmann, K., Rodan, I., Thayer, V., Tompkins, M. and Wolf, A. (2001) 2000 Report of the American Association of Feline Practitioners and the Academy of Feline Medicine Advisory Panel on Feline Vaccines. *J. Feline Medicine and Surgery* **3**, 3-10.

Liao, J.T. and Oehme, F.W. (1980) Literature reviews of phenolic compounds, IV, o-Phenylphenol. *Veterinary and Human Toxicology* **22**, 406-408.

O'Neill, L.L., Burkhard, M.J., Diehl, L.J. and Hoover, E.A. (1995) Vertical transmission of feline immunodeficiency virus. *AIDS Research and Human Retroviruses* **11**, 171-182.

Pedersen, N.C., Sato, R., Foley, J.E. and Poland, A.M. (2004) Common virus infections in cats, before and after being placed in shelters, with emphasis on feline enteric coronavirus. *J. Feline Medicine and Surgery* **6**, 83-8.

Robertson, S.A., Taylor, P.M. and Sear, J.W. (2003) Systemic uptake of buprenorphine by cats after oral mucosal administration. *Veterinary Record* **152**, 675-678.

Robinson, A., DeCann, K., Aitken, E., Gruffydd-Jones, T.J., Sparkes, A.H., Werret, G. and Harbour, D.A. (1998) Comparison of a rapid immunomigration test and ELISA for FIV antibody and FeLV antigen testing in cats. *Veterinary Record* **142**, 491-492.

Rochlitz, I. (2003) Study of factors that may predispose domestic cats to road traffic accidents: Part 1. *Veterinary Record* **153**, 549-553.

Schulz, R.D. and Conklin, S. (1998) The immune system and vaccines. *Compendium of Continuing Education* **20**, 5-18.

Slingsby, L.S. and Waterman-Pearson, A.E. (2000) Postoperative analgesia in the cat after ovariohysterectomy by use of carprofen, ketoprofen, meloxicam or tolfenamic acid. *J. Small Animal Practice* **41**, 447-450.

Sparkes, A.H. (2003) Feline leukaemia virus and vaccination. *J. Feline Medicine and Surgery* **5**, 97-100.

Tasker, S., Helps, C.R., Day, M.J., Gruffydd-Jones, T.J. and Harbour, D.A. (2003) Use of real-time PCR to detect and quantify Mycoplasma haemofelis and "Candidatus Mycoplasma haemominutum" DNA. *J. Clinical Microbiology* **41**, 439-441.

Taylor, P.M., Robertson, S.A., Dixon, M.J., Ruprah, M., Sear, J.W., Lascelles, B.D., Waters, C. and Bloomfield, M. (2001) Morphine, pethidine and buprenorphine disposition in the cat. *J. Veterinary Pharmacology and Therapeutics* **24**, 391-398.

Chapter 9

NUTRITION AND WELFARE

Kit Sturgess and Karyl J. Hurley
Wey Referrals, 125-129 Chertsey Road, Woking, Surrey, GU21 5BP, UK
Waltham Centre for Pet Nutrition, 1 Freeby Lane, Waltham-on-the-wolds, Melton Mowbray, Leicester LE14 4RT, UK

Abstract: Nutrition has a crucial role in determining the health and welfare of an animal. Owners have a responsibility to ensure that their cats receive a nutritionally complete and safe diet. The food should be offered in accordance with their natural feeding behaviour and physiology. Cats are obligate carnivores, with significantly different nutritional requirements from dogs and humans. As a result of their specialisation to a carnivorous diet, cats have much more specific nutritional needs and a narrower range of tolerance for various dietary components, making dietary deficiency and toxicity more common. The particular qualities of food products to meet the requirements of kittens, adults and geriatric cats are considered. Appropriate and specific nutritional prevention and management of disease has been a major development in veterinary medicine in recent years, and includes strategies to treat obesity, currently the most prevalent nutritional problem in pets in Western Europe and the United States.

1. INTRODUCTION

The keeping of pets is a widespread phenomenon that transcends nearly all national and cultural divides. The importance of relationships with pets is reflected in the substantial interest that we express in their health and welfare, and the major efforts that have been devoted to understanding and improving these issues. Improvements in feeding practices and nutrition have had a major impact on the health and longevity of cats. During the second half of the 20th Century, advancements in understanding their nutritional requirements led to the virtual elimination of diseases associated with nutritional deficiencies. Discoveries emerging now hold the promise of allowing our pets to live longer and healthier lives, notably by preventing or delaying the progression of diseases, slowing the aging process itself, and modifying the clinical manifestations of disease.

1.1 Cat's Relationship with Man

The domestic cat (*Felis silvestris catus*) is a member of the phylogenetic order Carnivora, an order renowned for its aggressive and efficient hunters. Cats are obligate carnivores and, while it might seem incongruous that we now share our homes with such fearsome predators, it was undoubtedly the predatory nature of ancestral cats that started our long association with this species. The process of domestication comprised two main phases, namely animal keeping and then selective control of breeding and the emergence of breed and colour types that we recognise today (Thorne 1992).

Cats were initially attracted to human settlements by the abundance of rodent pests that were associated with their grain stores (see Chapter 1). Humans quickly recognised their 'pest control' value, selecting wild cats that were more tolerant of man, and began to promote and develop the relationship. This process is believed to have begun soon after the development of settled agricultural communities in Egypt some 5,000 years ago (Serpell 2000) (although recent evidence suggests that cats may have started to associate with humans 9,500 years ago; see Chapter 3). Domestic cats progressively appeared in other countries from 500BC onwards, so that by AD1000 they were widespread throughout Europe and Asia. In the 17th century, they were taken to North America, and later on to Australia, in response to demands from settlers for help in controlling rodent populations. This contribution of cats to their relationship with man, namely their ability to control rodent populations, required no special training or modification of the cat's natural behaviour. For this reason there was little selective breeding of cats until the middle to late 19th century, when the breed and colour types of modern cats began to emerge.

The most common function of modern pet cats is that of companionship, where the rewards from keeping an animal are derived from the relationship itself rather than, for example, from help with vermin control. Dogs and cats are unique among domesticated animals in that their relationship with man exists without them being tethered or caged, and this, as well as their well developed social, sensory and communication skills, undoubtedly explains their popularity as pets. Pet ownership brings with it a number of benefits, namely the fulfillment of emotional or social needs, as well as improvements in human social relationships and psychological and physiological health (Robinson 1995; see Chapter 3).

1.2 The Importance of Diet and Nutrition

A slow process of mutual adaptation has driven the domestication of cats, where both humans and cats have gained more than they have lost during this process. Cats have sacrificed some of their freedom and been required to adapt certain of their natural behaviours, but in return they have been rewarded with food, shelter and protection. Domestication has inevitably been associated with a restriction in the degree to which cats can choose their own foods, meaning that we as owners are beholden to offer foods that are nutritionally complete and safe, and to offer these foods in ways that are sympathetic to the natural feeding behaviours and physiology of cats.

Early pet foods were formulated without regard to the nutritional requirements of cats, for these were not known at the time, and it was fairly common for the same foods to be sold to both dogs and cats, albeit in different packaging. It was not until the 1960s that efforts were made to understand the nutritional requirements of dogs and cats, and to regulate the production and sale of pet foods. At this time, the Association of American Feed Control Officials, which was formed in 1909 to regulate animal feeds, became involved with pet foods. The National Research Council, an organization formed to advise the United States government on scientific and technical issues, has, through its Committee on Animal Nutrition, published a series of documents entitled *Nutrient Requirements of Domestic Animals*. Their documents on the nutrient requirements of cats were first published in 1978, updated in 1986, and then comprehensively re-written and published in 2003; these are generally recognised as the standards for the manufacture of pet foods (National Research Council 2003).

The widespread availability of petfoods, that provide complete and balanced nutrient intakes, has led to the eradication of almost all diseases associated with nutrient deficiency or imbalance. There is, however, still progress to be made in harnessing the power of dietary factors to promote health and longevity, and to prevent the onset or slow the progress of certain diseases such as renal failure. There are also concerns over the possible role that diet plays in feline obesity and conditions such as diabetes mellitus and lower urinary tract disease. With increasing awareness that food should satisfy key behavioural as well as nutritional requirements of cats, the format of cat foods, as well as the way in which they are offered, are other important considerations. This chapter describes the nutrient requirements and feeding behaviour of cats, discusses the nature of the human-cat relationship as it relates to food, and presents guidelines on nutrition for optimum health and the management of specific disease states.

2. NUTRITIONAL REQUIREMENTS

Cats are obligate carnivores and therefore have significantly different nutritional requirements from dogs and humans. They have a much narrower range of tolerance for various dietary components; hence deficiencies and toxicity problems associated with nutrition are relatively more common. As the carnivorous specialization of cats developed, they lost their ability to synthesize a number of compounds in sufficient quantities. These compounds have now become essential dietary requirements. Hepatic function in cats has become specialized towards the metabolic demands of a carnivorous diet, and lacks the ability to metabolize and excrete a number of compounds, particularly those that require glucuronidation within the liver. Feline red blood cells are more sensitive than those of other species to the oxidant intoxicants, such as onions, that can be present in some foods.

2.1 Proteins

Cats have a higher dietary requirement for protein than dogs or man (Dickinson & Scott 1956), as there is an essential demand for protein to be used for gluconeogenesis regardless of the amount of carbohydrate in the diet. Gluconeogenesis from protein will continue to occur, albeit at a reduced level, despite protein malnutrition and negative nitrogen balance. The dietary requirement for protein varies with life stage but for an adult cat is set at 26% by the Association of American Feed Control Officials (AAFCO). Further constraints are placed on the protein source, as cats have a high dietary requirement for taurine and arginine.

2.1.1 Taurine

Unlike dogs, cats have an absolute requirement for taurine (ß-sulphonic amino acid [2-aminoethane sulphonic acid]). They are unable to manufacture sufficient taurine from the amino acids methionine and cysteine to meet their needs, due to reduced enzyme activity. Taurine is only naturally present in foods of animal origin. A dietary deficiency in preformed taurine is further exacerbated by the cat's increased use of taurine, relative to glycine, to manufacture conjugated bile salts. Cats almost exclusively use conjugate bile (cholic acid) to emulsify dietary fats prior to digestion and absorption. In the past, taurine deficiencies have occurred as cats' essential requirement was not recognized. Current recommended taurine levels in manufactured diets are 1500 ppm in dry versus 2500 ppm in canned food. The increased requirements in canned food compensate for losses during processing (Hickman *et al.* 1990). Taurine deficiency is now relatively rare, and is most commonly encountered in cats fed dog foods or vegetarian diets. Taurine deficiency can cause a variety of diseases: the most easily recognized are retinal degeneration (Barnett & Burger 1980), dilated cardiomyopathy (Pion *et al.* 1987) and reproductive failure (Sturman *et al.* 1986). Taurine deficiency is exacerbated by low potassium and acidifying diets. Taurine requirement also depends, to some extent, on the level of other sulfur-containing amino acids within the diet.

2.1.2   Arginine

Cats also have a higher requirement for arginine, a key intermediate in the urea cycle, than many other species. Cats fed a diet without arginine became encephalopathic within hours (Morris & Rogers 1978). Under many circumstances, arginine becomes a conditionally-essential amino acid. Diets such as baby foods are arginine deficient for cats, as are some of the liquid veterinary foods designed for use in dogs. Kittens require a minimum of 1.25% and adult cats 1.04% of arginine in their diet on a dry matter basis (DMB) (Anderson *et al.* 1979).

2.2   Fats

Cats have a reduced ability to synthesize some fats. In general, cat food has a slightly higher fat content than dog food. Apart from obesity, no diseases associated with feeding cats high fat, nutritionally balanced diets long-term are recognized. Fat-deficient diets can cause problems due to the lack of total energy, essential fatty acids (EFAs) and fat-soluble vitamins (A, D, E and K). EFAs are required to support many metabolic processes and are precursors for steroid hormone synthesis. EFA requirements for cats are difficult to estimate but a combination of linoleic and arachidonic acid is required. Current AAFCO recommendations are 0.5% linoleic acid and 0.02% arachidonic acid in diets containing 4000kcal of metabolisable energy (ME)/kg. Although beneficial in many ways, high levels of n-3 polyunsaturated fatty acids (PUFAs) can interfere with arachidonic acid absorption. Unlike dogs, cats are unable to convert linoleic acid into arachidonic acid, so arachidonic acid is listed as an essential amino acid in cats (MacDonald *et al.* 1984). Lack of arachidonic acid is associated with infertility in queens. Although not proven, a source of n-3 PUFAs is likely to be essential for normal metabolism and reproduction in cats.

2.3 Carbohydrates

Cats have a unique carbohydrate metabolism. There is an essential requirement for some gluconeogenesis to occur from protein metabolism, regardless of the amount of carbohydrate present. Indeed, amino acids such as arginine are more potent secretagogues of insulin than glucose (Kitamura et al. 1992). A cat's natural diet contains low levels of carbohydrate (there is about 3% in a mouse). This has led to a reduced capacity to metabolize carbohydrate, associated with decreased amylase production from the pancreas (about a third that of the dog) and lower levels of diassacharidases. It has also allowed the cat to develop altered hepatic metabolic pathways for the conversion of amino acids, such as serine, into glucose. High carbohydrate diets in cats can be associated with osmotic and secretory diarrhoea, as undigested carbohydrate passes into the lower small and large intestine undergoing bacterial fermentation. Following eating, the postprandial rise in blood glucose is significantly less in cats than dogs.

2.3.1 Fiber

Fiber refers to a group of complex carbohydrates that are resistant to enzymic digestion in the small intestine. Fiber usually undergoes bacterial fermentation in the large intestine and is classified on the basis of ferementability and solubility.

Although fiber is not an essential nutrient *per se*, there is good evidence to support its role in gastrointestinal health, colonic weight (Scheppach 1994) and the balance of commensal bacterial flora (Sparkes *et al.* 1998). Insoluble fiber primarily serves to increase bulk and maintain normal gastric transit time. Soluble fiber is fermented to short chain fatty acids that are important energy sources for colonocytes. High fiber diets tend to have reduced palatability for cats. In general, the fiber content of feline diets is lower than that in canine diets.

2.4 Vitamins, Anti-oxidants and Minerals

Vitamins are essential for health and well-being. They are generally classed as water (B and C) or fat (A, D, E and K) soluble. Vitamin C is not an essential part of a cat's diet, as cats are able to produce sufficient amounts in the liver.

2.4.1 Vitamin A

Vitamin A is only present in animal tissue. Most species can convert the precursors (carotenes) in plants to vitamin A, but cats lack the dioxygenase enzyme necessary and therefore require a preformed source of vitamin A in their diet. The minimum requirement for vitamin A for growth is 6000 IU/kg (DMB), which is approximately double the amount required for maintenance. Cats also transport vitamin A in an unusual form, as retinyl esters, rather than depending on retinal-binding protein (Raila *et al.* 2001). Despite the need for preformed vitamin A, deficiency is rare in cats except where they have been fed a vegetarian food. Affected cats are lame or show neurologic signs. Corneal vascularisation and ulceration, retinal degeneration, infertility, weight loss and muscular weakness are also reported.

Sporadic cases of vitamin A toxicosis still occur in cats, and are usually associated with the feeding of diets consisting almost entirely of liver or other organ meats. Such diets result in a deforming cervical spondylosis syndrome, due to the effects of vitamin A on bone growth. Initial clinical signs are of anorexia, weight loss and lethargy associated with pain and reluctance to move (Seawright & English 1967).

2.4.2 B Group Vitamins

B group vitamin deficiencies and toxicities are rare in cats. Diets are usually well supplemented, although the optimum levels of inclusion are largely unknown. Whilst not showing signs of overt deficiency, cats with low levels of vitamin B_{12} and gastrointestinal disease show a poorer response to treatment than cats with a normal level of B_{12}.

2.4.2.1 Niacin

Niacin (vitamin B_3) requirement is met in most species from dietary nicotinamide and through the conversion of tryptophan to nicotinic acid. Cats are unable to perform this conversion. However, as niacin is available in both animal and plant derivatives, the likelihood of niacin deficiency is very low.

2.4.2.2 Thiamin

Thiamin (vitamin B_1) is an essential co-enzyme in a number of metabolic processes. It is present in both animal and plant tissue with liver, whole grains and yeast being the richest sources. Thiamin is, however, very labile so the contribution of preformed thiamine in manufactured pet foods is very low; as a result thiamin is added as a supplement. Thiamin deficiency has been reported in cats and is usually associated with feeding raw fish and shellfish, which contain high levels of thiaminase. Clinical signs include anorexia, lethargy, neurologic signs, and muscle weakness manifested initially as ventroflexion of the neck (Davidson 1992).

2.4.3 Vitamin D

Like most animals, cats are able to synthesize vitamin D from 7-dehydrocholesterol when the skin is exposed to sunlight (UV radiation). However, the 7-dehydrocholesterol is more likely to be converted into cholesterol, making it most probable that cats need a source of preformed vitamin D in the diet (Morris 1996). Vitamin D deficiency and the development of rickets or osteomalacia is, however, rare and usually associated with an imbalance of calcium and phosphorus (see later). Rickets is seen as bowing of the long bones and thickening of the joints, associated with axial and radial thickening of growth plates and cupping of adjacent metaphyses. The distal radial growth plates are usually the most severely affected. Vitamin D toxicosis occurs almost exclusively as a consequence of over-supplementation and results in soft tissue calcification, although over-supplementation for prolonged periods is usually necessary for clinical signs to occur (Morita *et al.* 1995).

2.4.4 Vitamin E

Vitamin E is an important group of anti-oxidants that functions synergistically with selenium. Alpha-tocopherol is the most biologically active form of vitamin E. Requirements for vitamin E depend on the PUFA content of the diet. Naturally occurring deficiency is rare except where dry diets have been poorly stored and the fat has become rancid. Clear evidence of vitamin E toxicity has not been demonstrated.

Pansteatitis (yellow fat disease), associated with marginal levels of alpha-tocopherol and high levels of PUFAs, has been reported in cats (Tidholm *et al.* 1996). Clinical signs include anorexia, depression, pyrexia and pain and/or hyperaesthesia over the thorax and abdomen; coat condition may be poor. Cats are reluctant to move and palpation, which reveals granular or nodular subcutaneous fat deposits, is painful. Diagnosis is based on biopsy of the affected fat. Cases are most common in cats fed canned fish-based foods.

2.4.5 Anti-oxidants

Much of the initial effort in feline nutrition was directed towards ensuring diets met minimum nutritional requirements. More recently, the focus has shifted towards the role of diets in maximizing health and reducing the occurrence of disease. Key components in this area are anti-oxidants. Antioxidants are divided into two main categories, naturally occurring and synthetic. Naturally occurring anti-oxidants are principally the vitamins A and E and inorganic compounds such as zinc, copper and selenium. A variety of other compounds are also of interest, including bilberry, rosemary and gingko biloba.

Anti-oxidants have a central role in mopping up free radicals (reactive oxygen species) that are produced by various metabolic processes. Reducing levels of free radicals limits oxidative cell damage, leading to improved health. Large scale, human studies looking at dietary anti-oxidants in the prevention of cataracts have shown a reduced risk of cataract development in individuals taking vitamin supplements (Mares-Perlman *et al.* 2000) (the development of cataracts is thought to be strongly associated with oxidant damage to the lens tissue). This effect was most notable when compared with a group who were receiving poor quality diets.

Synthetic antioxidants are most commonly butylated hydroxytoluene or hydroxyanisole or ethoxyquin. They are used to prevent auto-oxidation of processed diets, as they better withstand the heat, pressure and moisture during food processing.

2.4.6 Minerals

Minerals are the inorganic components of food. More than 18 minerals are believed to be essential and are divided into three major classes, macrominerals, microminerals and ultra-trace minerals, according to the quantity that is required in the diet (Table 1). Whilst deficiency in any of these components is possible, abnormalities of potassium and phosphorus or calcium balance are most commonly encountered. Variable intake of iodine may be a factor in the pathogenesis of hyperthyroidism in cats (Gerber *et al.* 1994).

2.4.6.1 Calcium deficiency and phosphorus excess

This is the most common mineral-associated disease seen in cats, and is usually associated with feeding an all-meat diet resulting in low calcium and high phosphorus intake (calcium to phosphorus ratio $< 1:16$), which causes a nutritional secondary hyperparathyroidism. Kittens are at greater risk than adult cats due to their increased calcium requirement. Clinical signs include lameness; a reluctance to stand; pain; enlargement of costochondral junctions and metaphyses; pathological fractures leading to angular limb deformities or neurologic signs associated with vertebral fractures; constipation; and hypocalcaemic cataracts. The diagnosis is based on biochemical analyses (although not all cases have low serum calcium) and radiographic changes, which show areas of relative osteopaenia in metaphyses, especially the distal radius and ulna, thin cortices, bowed diaphyses and pathological fractures. The osteopaenia is best appreciated in the dorsal spinous processes of the thoracic vertebrae (Tomsa et al. 1999).

Table 1. The main minerals in food and their functions in the body

Class of mineral	Functions in the body
Macrominerals	
Calcium	Major skeletal component, blood clotting, cellular movement of calcium, essential for excitable cell function (nerve, muscle) and membrane permeability
Phosphorus	Bone and muscle constituent, metabolism, energy production, reproduction
Sodium/chloride	Osmotic and acid-base balance, nutrient, waste product and water metabolism, nerve transmission
Potassium	Excitable tissue function, acid-base and osmotic balance, enzyme co-factor
Magnesium	Bone and intracellular fluid, excitable tissue function, enzyme component, carbohydrate and lipid metabolism
Microminerals	
Iron	Enzyme constituent, haem proteins, activation of oxygen
Zinc	Major enzyme constituent/activator, anti-oxidant, skin integrity, wound healing, immune response, foetal development and growth
Copper	Enzyme systems, haemoglobin formation, connective tissue, myelin and bone formation, pigment, cardiac and immune function
Manganese	Enzyme factor in carbohydrate and lipid metabolism, Bone development, reproduction, cell membrane integrity
Iodine	Component of thyroid hormones
Selenium	Component of glutathione peroxidase (a major antioxidant pathway), immune and reproductive function
Ultra-trace	
Chromium	Potentiates insulin activity
Boron	Calcium and phosphorus metabolism via PTH, magnesium and cholecalciferol metabolism

2.4.6.2 Potassium

Cats appear poorly able to regulate their potassium metabolism, and rapidly become hypokalaemic if they stop eating. A breed-associated hypokalaemic polymyopathy is described in Burmese cats. Potassium requirement is dependent on the protein content and acid load of the diet. Normal cats fed low potassium, acidifying diets have been shown to develop renal failure (DiBartola *et al.* 1993).

2.5 Common Problems Associated with Food

Due to their unique metabolism, there are a number of foods that should not form a major part of a cat's diet. In general, problems are not likely to occur if such foods constitute less than 10% of the dietary intake.

2.5.1 Meat and Dairy Products

Cats are carnivores but cannot survive on just lean meat, which has excessive amounts of phosphorus relative to calcium and is deficient in sodium, iron, copper, iodine and vitamins. An all-lean meat diet can lead to severe and potential fatal skeletal abnormalities, joint malformations as well as essential fatty acid deficiencies and, classically, a nutritional secondary hyperparathyroidism. Liver (and other organs) contains excessively high levels of vitamin A, leading to painful bone deformities which do not resolve, even if the diet is corrected.

Milk products contain large quantities of fermentable sugars that require lactase for metabolism. Like most species, lactase levels decline as kittens mature. If there is insufficient lactase present in the gut, the fermentable sugars will reach the colon leading to osmotic diarrhoea.

2.5.2 Fish

Raw fish can contain the enzyme thiaminase, which destroys the vitamin thiamine (B_1), and may also contain parasites. Excessive amounts of fish, especially if packed in oil, can cause a relative deficiency of vitamin E leading to pansteatitis. Cheap tinned fish, especially tuna, can contain preformed histamine, resulting in vomiting and diarrhoea.

2.5.3 Vegetarian Diets

Cats are obligate carnivores and require meat in their diet for optimal health. Vegetarian diets have to be supplemented with taurine, vitamin A and arachidonic acid to be nutritionally balanced. Even so, the general health status of cats fed supplemented vegetarian diets is not as good as those fed meat based diets.

2.5.4 Other Foods to Avoid

Baby food can often be arginine deficient for cats. The baby food may contain onion powder that can cause an oxidative anaemia.

Dog food is not nutritionally balanced for cats and can have inadequate amounts of taurine, vitamin A and arachidonic acid, especially if it is a cereal-based diet.

3. FEEDING BEHAVIOUR

The feeding behaviour of cats encompasses many facets, from the search for food or prey through to its acquisition, consumption and post-prandial activities like grooming and sleeping. The wild ancestors of domestic cats had sophisticated strategies for locating food and then determining whether it was safe to eat and likely to meet their dietary needs, and many of these behaviours are identifiable in today's pets (Bradshaw & Thorne 1992).

3.1 Feeding Behaviour of Wild Cats

Wild cats face enormous challenges in locating and catching prey and then working out whether the food is safe to eat and nutritionally adequate. The feeding behaviour of feral cats has been extensively studied: their diet is almost exclusively carnivorous, comprising primarily small mammals with some birds, reptiles and insects, with the catch determined by what is locally available and by seasonal variations. Rats and mice are often caught but not always eaten, suggesting that they are less palatable than the preferred prey of voles and young rabbits and hares. Insectivorous mammals, such as shrews, are probably the least palatable prey because they are rarely eaten when caught (Bradshaw & Thorne 1992).

Cats use their senses of sight and hearing for detecting prey; sound initially attracts a cat's attention but it is the motion of a prey-sized object that is primarily responsible for initiating the hunting sequence. Two main hunting strategies have been observed, and these compromise either a slow stalk or waiting to ambush the prey (Fitzgerald & Turner 2000). In the former, the cat walks slowly, looking and listening, and stops periodically to concentrate upon some sight or sound. In the latter, the cat identifies a suitable target, for example the entrance to a rodent burrow, and waits, keeping its body still while staring at the entrance. Once prey appears, the cat usually waits for it to move away from its refuge before pouncing. Cats usually kill with a bite to the neck, severing the spinal cord, and use their forepaws to hold larger prey or dislodge the prey from a site of refuge. The prey is either consumed close to the point of kill or carried home, either dead or alive, where it may be 'played' with before being killed and eaten by the hunter or other cats. Some cats seem to specialise in hunting birds and develop skills to overcome the vigilance of birds and their wide field of view; these include a rapid dash to hide under cover close to prey and creeping towards the bird using as much cover as possible. This said, many cats are actually unsuccessful at catching adult birds and prey predominantly on young, whereas others learn to catch insects with their forepaws.

Cats are born with innate predatory abilities, and this is apparent when cats, that have never had to hunt to survive, demonstrate predatory behaviour when presented with an appropriate stimulus (Robinson 1992). Effective hunting skills do, however, require practice and this occurs at an early age, with feral mothers bringing prey to their kittens at around 4 weeks of age for their consumption. Kittens are then encouraged to investigate the prey in the nest, with the dam demonstrating how to kill and consume the food. By observing what food their mother eats, the kittens learn what is safe to eat and establish a strong drive to consume the same foods (Bateson 2000).

Because feral cats feed primarily on small mammals, they need to make several kills each day to obtain sufficient food and this is an important factor in determining how we feed domestic cats. Patterns of hunting and feeding in wild carnivores are related to the availability of food, namely the abundance and size of prey, and are modified if the cat has access to other foods, for instance those supplied by humans. Cats regulate the amount that they eat to maintain a set weight, and the pattern of taking frequent small meals is modified when prey becomes scarcer: the frequency of feeding decreases and meal size increases, resulting in a more or less constant total food intake.

3.2   Food Selection in Domestic Cats

Despite restriction of their freedom to chose foods, domestic cats still display considerable flexibility in their food choices. Several factors are important in determining food selection, namely palatability, nutritional content and previous experience (Bradshaw & Thorne 1992). The term palatability is used to describe the attraction of a foodstuff and relates to its sensory properties, that is appearance, smell, taste and texture. Palatability does not appear to play a major role in determining food selection in wild cats, but may be important in forming preferences for a particular prey species; the same behavioural mechanisms are thought to underlie food selection in pet cats. Of the sensory aspects of food, it is likely that taste has the most significant bearing on palatability in cats.

The sense of taste in cats is communicated via taste buds located on the upper surface and at the back of the tongue and on the palate (Bradshaw 1992). Certain taste bud receptors are able to detect chemicals that accumulate in mammalian tissues after death, thus measuring the freshness of food, and may explain the cat's aversion to carrion (Bradshaw 1992). Other receptors are stimulated by sulphur-containing amino acids, which include taurine, providing evidence of a link between the palatability of a food and the requirement for this essential amino acid in cats (see section 2.1.1). Cats do not demonstrate a great attraction to sweet foods, which is believed to reflect their adaptation to a meat-based diet.

The palatability of meats is thought to be related to the species of origin and the type of tissue from which the flesh is taken, as well as its freshness and the nutritional status of the animal at the time of its death or slaughter. Processing methods and the interaction of different ingredients during the cooking process also have an impact on palatability. Cats are popularly believed to prefer fish to meat. Individuality in food preferences exists, with some cats described as finicky eaters. While the basis for individual food preferences is unknown, it is likely that genetic factors and early food experiences are key.

With the exception of salt, there is no evidence that cats can recognise missing nutrients by taste, nor are they capable of immediately recognizing a nutritionally deficient or imbalanced diet. Feeding a nutritionally inadequate diet does, however, result in declining food intake or complete refusal to eat that diet. This is believed to represent a learned aversion, where the sensory characteristics of the food, notably its taste, are associated with the malaise that follows its consumption (Bradshaw & Thorne 1992). This process can conceivably be subject to errors of interpretation, for example where a cat erroneously associates the symptoms of a gastrointestinal infection with the taste of the previous meal, causing it to avoid the flavours in that meal (see section 5.2).

When cats are faced with a new food, for which they have not assimilated information on its safety or nutritional adequacy, they adopt several strategies. One of these is neophobia, where the food is first sampled and then, if found to be safe, is eaten in greater quantities on subsequent occasions (Hill 1978). In some extreme cases, a new food may be repeatedly rejected and such animals are described as being food-fixated. The experience of monotony can also alter feeding preferences, with prolonged feeding of a single food leading to transitory preferences for other foods, even those less palatable than the staple diet (Thorne 1982). This adds credence to the idea that dietary variety is important for cats.

3.3 Patterns of Food Intake in Domestic Cats

Cats naturally regulate their food intake so that they do not gain excessive weight, and the most important regulator of body weight in meal-fed pets is the size of the meal, which is in turn controlled by satiation. This is where the end of the meal is signaled in anticipation that sufficient food has been taken into the stomach to meet the animal's nutritional requirements. This process is only accurate as long as the end products of that food's digestion can be predicted, which generally means that the food is a familiar one. When presented with a new food, particularly one of a novel format such as a semi-moist food, cats can over- or under-eat until the link between the amount of food consumed and its yield of energy and nutrients can be established (Bradshaw & Thorne 1992).

Pet cats have relatively little control over their food supply, although they may beg or harass their owners into providing food out-with mealtimes. When a cat has free access to food, it takes small meals throughout the day and night, with the number of meals ranging from 7 to 16 in any 24-hour period (Mugford & Thorne 1980). In domestic situations, cats are usually offered two or more meals a day and appear to adapt well to these enforced patterns (Figure 1). Even when the number of meals offered is small and finite, they do not usually eat all the food immediately, thereby increasing the number of meals taken throughout the day and mimicking the behaviour of wild cats. Some cats, however, are less well able at adapting to being given one or two large meals a day, consuming the food rapidly and completely, and often prompting their owners to offer more food. This type of behaviour can lead to over-consumption of energy and the development of obesity, so it is important in these situations to limit the amount of food offered rather than providing it on an *ad libitum* basis.

The rate at which food is consumed is a function of its moisture content, with wetter foods being more easily handled and more rapidly consumed (Mugford 1977). Moist foods are eaten rapidly at the start of a meal but the rate then declines, whereas dry foods are consumed at a slower and more constant rate. These patterns are also influenced by the texture of the food and its palatability, with textural variation improving the ease with which food is handled and foods of lower palatability being consumed more slowly.

Figure 1. When a cat has free access to food, it takes small meals throughout the day and night. In domestic situations cats are usually offered two or more meals a day and appear to adapt well to these enforced feeding patterns.

In the home, other influences on feeding patterns can become apparent. The most obvious of these is the role that food plays in the relationship between the pet and its owner and particularly in reinforcing behaviour. Cats are adept at learning behavioural sequences, such as begging, that result in food being given to placate the animal or used as a reward. Food may also be left as a palliative, for example to compensate for the time that cats are left alone. Each of these factors can result in excessive energy intake and lead to obesity. Inappetence is very much less common, although it does occur in cats, and can be overcome by mimicking natural feeding patterns and offering the food at body temperature, which increases its palatability. Cats may well eat less when the ambient temperature is high and more when it is cool, and changes in the environment, such as the absence of the owner or being in unfamiliar surroundings, can also influence appetite.

Despite the provision of food by their owners, many cats continue to hunt, which they do during daylight as well as under the cover of darkness. Cats frequently bring prey back to the household that may be dead or injured, and this is thought to reflect an incomplete sequence of food caching in which, if left undisturbed, the prey is often retrieved and consumed over the next few days.

Cats have extraordinarily efficient kidneys – a reflection of their desert origins – and demonstrate almost a reluctance to drink water (see section 4.3.2). It would appear that practically all of their water requirements can be met from water contained in moist foods; cats fed canned foods may rarely drink. Cats fed dry foods are stimulated to drink water by the inclusion of about 3% salt in the ration, as a precaution against the development of lower urinary tract disease.

4. NUTRITION FOR OPTIMAL HEALTH

Most food products for cats, apart from treats and snacks, contain a statement of nutritional adequacy and generally fit into one of four categories. The first category covers all life stages, the second is restricted to a specific life stage (e.g. kitten, junior, adult, senior), the third includes diets suitable for intermittent or supplementary feeding (i.e. they are not nutritionally adequate for sole, long-term feeding), and the final category consists of therapeutic diets (diets that are nutritionally adequate only for specific conditions, and used under veterinary guidance).

4.1 Nutrition for Kittens

4.1.1 Pre-weaning Nutrition

The first few hours of life are critical to the health of a kitten, as this is the time that passive immunity is transferred via the colostrum. Colostrum contains immunoglobulins, large proteins that would normally be digested but initially are absorbed intact across the intestinal lumen. Less well characterized is the transfer of a cellular component and other tropic factors. Closure of this pathway occurs rapidly, and is probably associated with rising insulin levels (Donovan & Odle 1994); closure is completed by about 16 hours in kittens (Casal *et al.* 1996). After this time, immunoglobulins in the milk still have value, as they provide local immunity within the intestinal tract and are a concentrated source of nutrients, but can no longer contribute to systemic immunity. Colostrum also has an important function in establishing circulating blood volume (Fisher 1982). Newborn kittens are unable to concentrate their urine, so they have a high fluid requirement at around 120ml/kg/day.

The composition of cat milk changes during lactation, but these changes are relatively small compared to some other species (Adkins *et al.* 1997). After four weeks of age, milk alone no longer provides adequate calories or nutrients, so weaning usually begins from three weeks of age.

4.1.2 Feeding Orphan Kittens

Feeding an orphan kitten is only part of the care that needs to be provided; maintaining a suitable physical and emotional environment is equally important. The ideal solution is to find a foster queen, but as these are rarely available, milk substitutes usually need to be given. Several commercial milk substitutes for kittens are available, and these are preferable to homemade preparations. Many of the homemade recipes available provide a nutrient composition that is significantly different from queen's milk. Some commercial milk substitutes are also not well formulated and have caused diarrhoea and cataracts (Remillard *et al.* 1993). Cow's milk alone contains too few calories as protein and fats, and an excessive level of lactose and casein for neonatal kittens (Lepine 1998); the milk is also significantly more dilute. While most commercial supplements are based on cow's milk, they are able to promote growth rates that closely match those of nursing kittens. Kittens require approximately 25% of their body weight per day in reconstituted milk replacer. However, amounts will vary with the supplement used and should be based on daily weight increases. Fresh formula should be made for each feed and fed at approximately 37.5°C (100°F). At birth, kittens will need to be fed every four hours; this can be reduced to six- hourly feeding by 3 to 4 weeks of age. Bottle-feeding should be performed with the kitten's head tilted upward to minimize air intake. Stomach tubing is quicker and reduces the risk of aspiration; there is, however, a higher risk of over- or under-feeding. The emotional impact on kittens of not being able to suckle has not been evaluated.

4.1.3 Post-weaning Nutrition

Semi-solid food should be provided from 3 to 4 weeks of age, and most kittens will be fully weaned by 6 to 8 weeks. Cats are considered adults by 10 to 12 months of age, so it is appropriate to feed a kitten (or junior) diet up to this time. Kitten diets are formulated to meet the needs of growth and maintenance, and therefore have different nutritional balance from adult foods. In general, foods should be energy dense and have high digestibility, palatability and biological value.

4.2 Nutrition of the Adult Cat

Adult cats have less demanding nutritional requirements than kittens. The nutritional idiosyncrasies of cats have already been discussed in relationship to their evolution as obligate carnivores.

More recently, nutritional research has been directed beyond meeting the requirements for survival towards optimal and preventative nutrition. This has resulted in a marked increase in the inclusion of antioxidants in the diet, as well as the development of diets to promote oral health and maximize skin and coat quality.

4.2.1 Diets to Promote Oral Health

Dental disease is a major problem in older cats. The prevalence is very high (60% in cats older than 3 years) (Crossley 1991), with dental disease being one of the most common reasons for older cats to be presented for veterinary attention. Two major dental problems exist in cats: periodontal disease, which is usually associated with plaque accumulation, and feline odontoclastic resorptive lesions (FORLs). Currently available diets can make a significant contribution to, but will not maintain, oral health in the absence of tooth cleaning. Diets are aimed at altering the oral environment to discourage the proliferation of plaque forming bacteria, stimulate the flow of saliva, maintain healthy gingival tissue and remove plaque by mechanical abrasion. The latter is achieved by shape, texture and fiber alignment and thus is only really achievable with a dry diet or chew (Watson 1994). Dry food *per se* will reduce plaque or calculus formation but the effect is limited (Studer & Stapley 1973) as the kibble is usually broken by the tip of the tooth and has no abrasive action at the gingival margin.

FORLs have been associated with diets that are low in magnesium, and the feeding of non-commercial diets (Lund *et al.* 1998). It has been speculated that the use of acid sprays as coating for dry diets may be involved in demineralising the teeth. In a study by Zetner and Steurer (1992), cats with FORLs had a lower tooth surface pH, but the pH was not related to the diet fed.

Oral health can best be maintained with an appropriate diet, in conjunction with tooth brushing and/or the use of oral antiseptics such as chlorhexidine. Many cats, however, do not allow their teeth to be brushed by their owners unless they are used to this procedure from an early age.

4.2.2 Nutrition of Queens

Pregnant queens exhibit a linear increase in weight, beginning around the second week of gestation. Approximately 40% of the weight gained during pregnancy is lost immediately following parturition. The remaining 60% is lost during lactation, due to a loss of body fat that has been accumulated during pregnancy. By the end of gestation, a queen should be receiving 25 to 50% more food than she did at maintenance (before the pregnancy). During lactation, the queen may require 2 to 2.5 times more food than maintenance, depending on the size of the litter. Many cats are fed *ad libitum* and this is a convenient method of ensuring adequate nutrition during gestation and lactation.

A single diet can be formulated that is suitable for both gestation and lactation in queens; it should be highly digestible and energy dense. Protein quality is important, to provide essential amino acids for growth and development of the foetus. A protein content of 35% (DMB) allows for optimal weight gain during gestation (Piechota *et al.* 1995). Taurine is an essential nutrient for fertility and reproduction in cats, and low taurine diets will result in foetal death and deformity.

Fat content of gestation and lactation diets is usually increased to help achieve the necessary energy density; diets containing 21 to 27% fat are suitable. A dietary source of docosahexaenoic acid is required for normal retinal development of nursing kittens; milk concentrations parallel dietary intake. Low fat diets are associated with smaller litter size and increased kitten mortality (Olovson 1986).

No particular increase in vitamins and minerals is required, as the increased food consumption meets the increased needs. Copper deficiency in well-supplemented diets has been caused by high zinc levels that compete for absorption, and low copper availability due to the use of copper oxide (Morris & Rogers 1995).

4.3   Geriatric Nutrition

From a nutritional, behavioural and metabolic point of view, changes in function can be demonstrated in cats as early as 7 to 8 years. Overt signs of old age are usually seen by the time a cat is 10 to 12 years old. Many cats are now spending more than 50% of their lives being 'old'. The nutritional needs of older cats have been investigated with the aim of maintaining health, maximizing longevity and reducing the risks of developing age-associated disease. A number of studies have reported the effects of aging in cats (Markham & Hodgkins 1989; Taylor *et al.* 1995; Harper 1998). Whilst many of the physiological changes that occur with aging have been defined and suitable nutritional solutions sought, there is still no clear evidence that feeding an otherwise healthy, older cat a specially-formulated senior diet will affect health, longevity or disease occurrence. The only known nutritional intervention that has been shown to increase longevity, in mammals, is lifelong caloric restriction.

As intestinal function is reduced, smaller more frequent feeds are indicated and many older cats achieve better nutrient intake if allowed to 'graze', that is eat small meals little and often. Older cats should have their weight and food intake monitored regularly. The major changes associated with aging in cats are:

- Increased sensitivity to dehydration and reduced total body water content
- Reduced tolerance to heat or cold
- Decreased major organ function, including reduced nutrient digestion and absorption
- Dental disease
- Decreased salivation, taste sensation and olfaction
- Decreased adrenal gland function
- Decreased cardio-respiratory function
- Reduced activity and lean body mass, associated with a decreased basal metabolic rate, that results in loss of muscle mass and skeletal density with increased body fat
- Altered neurologic function
- Reduction in immune function

Some of these changes affect nutrient assimilation and incorporation (Taylor *et al.* 1995), whilst others may require the nutritional balance of the diet to be altered. Older cats have a decreased ability to adapt to change and may not tolerate sudden changes in diet, even if the new diet may have potentially beneficial effects in the long-term.

4.3.1 Weight

Whilst obesity is an increasing problem in cats, the percentage of cats that are overweight declines with increasing age and the number of underweight cats increases. An important nutritional goal is to try and maintain a stable and optimal body weight.

4.3.2 Water

Water is a critical factor, as an increase in fluid requirement is common in a number of diseases, such as renal dysfunction, that are prevalent in older cats. Fluid balance is more likely to be a problem in cats on dry food diets, as a cat weighing 4 kg needs to drink a minimum of 150 ml of water per day to meet its fluid needs. By comparison, a wet food diet will supply the majority, if not all, of its fluid needs. An adequate water balance can be a particular problem in an older cat if changed from a wet to a dry food diet.

4.3.3   Energy

Well-controlled studies are lacking, but it is widely believed that older cats require less energy (Taylor *et al.* 1995) and many geriatric diets have reduced energy density. Decreased energy requirement is associated with a lower resting metabolic rate and activity. Whilst the incidence of obesity declines with age, a significant number of cats over 8 years are still obese; the number of underweight cats begins to increase significantly from around 13 years of age. Both obesity and cachexia are associated with a significant increase in mortality in cats over 8 years of age.

The ability of older cats to digest fat is reduced, suggesting that energy may be better provided by carbohydrate. However, due to their limited ability to metabolize carbohydrates, excess in the diet can lead to diarrhoea.

4.3.4   Protein

Protein restriction in old cats is inappropriate except in the face of other disease, such as chronic renal or hepatic failure. As mentioned in section 2.1, cats have an absolute requirement for protein; higher protein diets also tend to be more palatable for most cats. Diets should contain 30 to 45% protein (DMB), which should be of high biological value (meat proteins).

4.3.5   Fat

Sufficient fat must be given to provide essential fatty acids and maintain energy density and palatability. Increased levels of n-3 PUFAs may be beneficial in the control of hypertension, maintenance of renal blood flow and in reducing the level of deleterious cytokines that may be involved in weight loss in older cats. However, excessive levels of PUFAs can lead to the accumulation of peroxidised fats, which increase free radical damage. Aging has also been associated with a decrease ability to desaturate fats, so the relative proportion of desaturated fats in the diet may need to be increased.

4.3.6 Calcium and Phosphorus

On the basis that many older cats have renal compromise, many diets are mildly phosphate-restricted. In order to maintain a balance, calcium levels in the diet are also reduced. As protein is the major source of dietary phosphorus, a degree of protein restriction is inevitable.

4.3.7 Potassium

Potassium requirement is a function of the protein level in the diet. Older cats may have an increased requirement due to chronic renal failure, where potassium wasting occurs secondary to polyuria and acidosis. Acidifying diets, such as those designed for the management of urolithiasis, increase potassium loss and are probably inappropriate in the majority of older cats. The risks of struvite urolithiasis are substantially reduced in older cats, who have an increased risk of oxalate urolithiasis.

Older cats may also require increases in dietary magnesium.

4.3.8 Vitamins

Vitamin demand may be higher in older cats, so most diets are supplemented. However, oversupplementation, particularly of vitamins A and D, can be harmful. Vitamin inclusion rates should be between 2 to 5 times the recommended daily allowance. Increased levels of anti-oxidants have also been proposed to support the declining immune system.

4.3.9 Cognitive Function

Many cats appear to have a decline in cognitive function with increasing age, becoming confused about their environment and losing some aspects of learned behaviour (Houpt & Beaver 1981) (see Chapter 4). The use of anti-oxidants and compounds that promote improved cerebral circulation may help to delay or alleviate these changes.

4.3.10 Hypertension

Hypertension is a relatively common problem in older cats, and is associated with diseases such as chronic renal failure and hyperthyroidism. Diets that have relatively low sodium content and increased levels of n-3 PUFAs will tend to decrease the risk of clinically significant hypertension.

5. NUTRITION IN THE MANAGEMENT OF DISEASE

It is without the scope of this text to present a detailed discussion of the role of nutrition in the management of disease, for which specific texts are available (Kelly & Wills 1996; Case *et al.* 2000; Hand *et al.* 2000). The general approaches to the nutritional management of obesity, and to feeding sick cats, are presented below. Because cats are obligate carnivores and have more rigorous dietary requirements, the development of nutritional strategies for the management of disease has been more difficult. As a group, cats are less food orientated than dogs, and are therefore often less willing to accept manipulation of their diet. Some cats will show very strong food preferences, making dietary change very difficult; a gradual food change over two weeks or more may be necessary. Food intake should be maintained above 70% of normal, and care taken that the preferred food is not picked out of the mix. Trying to force a cat to eat a new food by providing no alternative is rarely successful, and can result in hepatic lipidosis (a serious, potentially fatal liver disease).

5.1 Obesity

Obesity (defined as greater than 15% above ideal weight) is the most prevalent nutritional problem in pets in Western Europe and the United States, and affects between 10 and 20% of pet cats (Figure 2). Rarely is obesity associated with enlargement of abdominal structures, metabolic or endocrine disease. However, a range of medical conditions should be ruled out before a weight loss program is started.

In the vast majority of cases, weight gain is associated with overnutrition, due to increased caloric intake or reduced requirement (for example due to less exercise). The problems associated with obesity include (adapted from Hand *et al.* 2000):

- respiratory difficulties
- decreased cardiac reserve
- insulin resistance and the development of diabetes
- poor response to infectious diseases
- fatty infiltration of the liver
- increased surgical risk due to increased risk of anaesthesia, fat necrosis, slower wound healing, technical difficulty in performing the surgery
- feline lower urinary tract disease

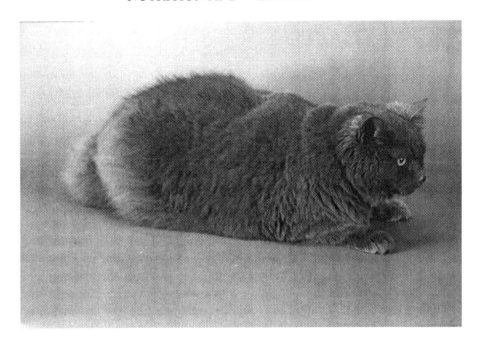

Figure 2. Obesity affects between 10 and 20% of pet cats in Western Europe and the United States.

Weight reduction is centered on reducing caloric intake, increasing exercise and other behavioural modifications of the cat and its owner (see Chapter 4). In general, there are few risks and complications associated with weight loss in healthy adult cats. In most instances, using a manufactured, calorie controlled diet is the easiest way to achieve weight loss, although the owner can try reducing the amount of the cat's normal food. The owner should also eliminate all high calorie snacks and treats, and ensure the cat is not getting food elsewhere. Increasing exercise should not be ignored, as even 10 minutes of daily exercise with a toy can be effective in promoting weight loss.

There is considerable variation in the nutritional content of proprietary weight reduction diets. In most diets, the caloric density is reduced by increasing the water or fiber content, providing calories as carbohydrates (as opposed to fats) or expanding the kibble volume. The ability to increase fiber and reduce fat in feline diets is limited by nutritional needs and palatability.

In the majority of cases, failure of the cat to lose weight is due to poor owner compliance so veterinary support, to encourage the owner to follow dietary recommendations, is crucial to success. As the caloric intake is reduced, a cat's basal metabolic rate (BMR) will fall, thus reducing its caloric requirement. This reduction in BMR can be partially prevented by increasing the cat's exercise. Because health benefits such as increased activity may lag behind weight loss, owners frequently become discouraged and give up the diet in the early stages, especially if the cat is pestering them for food.

5.2 Feeding the Sick Cat

Most nutritional therapies are not immediate in their effect, and consideration needs to be given as to the appropriate time to introduce a cat to a new diet. Feeding a new food when a cat is unwell, particularly if it is nauseous, or feeding a new food that causes diarrhoea or vomiting, can result in learned taste aversion and result in poor long-term acceptance of an appropriate diet. Foods that are strongly odorous or have high protein contents are more likely to induce aversion. Aversion has been documented to last for up to 40 days, but in the author's experience can appear to be permanent in some cats. For most cats, the key is to get them eating and feeling better first, and then introduce a new diet. If a cat has not eaten for more than three days, its nutritional needs should be addressed as a matter of urgency.

Critical patients are catabolic despite starvation, so they will lose body weight extremely rapidly. The lack of enteral nutrition causes reduced epithelial cell turnover within the intestinal tract and increased bacterial translocation from the intestine into the systemic circulation. Poor nutritional status is also associated with decreased immune function, delayed wound healing, increased risk of sepsis, prolonged hospitalization, and, ultimately, poor nutritional status may affect the prognosis.

5.2.1 The hospitalized patient

In general, hospitalized cats should be offered highly palatable, highly digestible, nutrient-dense foods in small quantities. Frequent feeding (4 to 6 times per day) is essential for critically ill or anorexic patients. There is probably nothing more unappetizing for a hospitalized cat than to be presented with a large bowl of food, that is left to go stale in the kennel and from which it cannot get away, so uneaten food should be removed 20 to 30 minutes after being offered. Some cats, however, will eat much better at night, or will graze very small quantities of food. Careful observation of its eating habits, and owner history, are important when devising a feeding plan for the anorexic hospitalized cat.

A number of supplements to standard diets should be considered, as these may improve outcome. Glutamine is a major substrate for gluconeogenesis and is the principal amino acid in plasma and muscles. It is also important for the normal stress response of many tissues, such as the kidney, white blood cells and fibroblasts. Glutamine supplementation of the diet has been shown to improve recovery rates (Klimberg & Souba 1990). Supplementation of the diet with branched chain amino acids may be advantageous in septic cats with encephalopathy. Adequate levels of arginine are essential for normal functioning of the urea cycle in cats, and arginine supplementation may promote wound healing and immune function (Kirk & Barbul 1990).

5.2.2 How much and how to feed

Nutritional calculations should be based on the resting energy requirement of the cat. On the first day of feeding after a period of starvation, between a third and a half of the calculated caloric need should be given, and the daily amount increased to the calculated amount over two to three days. The patient should be weighed daily.

Many cats will eat voluntarily, especially if given highly palatable, warmed, odorous food. Time should be taken to encourage the patient to eat, by sitting with it and offering food by hand. Sometimes inviting the owner to the veterinary surgery to feed their cat can have positive benefits.

If encouraging voluntary intake proves ineffective, alternative methods should be considered. Chemical appetite stimulants can be of value in the short term, but are only appropriate when the underlying problem, causing the inappetence, is being resolved. Force-feeding is counterproductive as this only increases the cat's aversion to food and is stressful for the patient and veterinary staff.

For the majority of patients, tube feeding is the next appropriate step to ensure adequate caloric intake. A variety of methods are available; the choice depends on the clinical problems of the patient, in terms of the disease process, and the length of time that nutritional support is likely to be required. The simplest method is a naso-oesophageal tube that can be placed in a conscious cat without the need for expensive equipment. It is suitable for short-term feeding (5 to 7 days), and its tube diameter is relatively narrow. Recently, oesophagostomy tubes are being more widely used; although anaesthesia is required, they also do not require specialized equipment for placement and can be maintained for relatively long periods. In cats requiring long-term support, or those with oesophageal disease, gastrostomy tubes fitted surgically, or preferably endoscopically, can be maintained indefinitely. They are generally well tolerated and of reasonable tube diameter, allowing medication as well as food to be delivered directly into the stomach. Jejunostomy tubes are normally fitted surgically where gastric bypass is necessary, but are difficult to maintain and of limited use in the long-term.

Total parenteral nutrition may be required in extreme cases. It is a complex and potentially risky method of nutritional support, as no nutritionally balanced preparations, suitable for long-term use in cats, are available. Due to their high osmolarity, feeding solutions need to be given via a jugular catheter.

6. CONCLUSIONS

Those caring for cats have a duty to keep them in good health and at optimal bodyweight by feeding them nutritionally balanced, palatable foods in the correct amount and in the appropriate manner. The formulations of cat foods are continually evolving as the knowledge base increases. The focus at present is on optimizing the health benefits at different life stages, and preventing or managing a range of diseases. In the future, it is likely that more individualized diets will become available, which are tailored to the specific needs of particular cats. Breed-specific diets, that optimize kibble shape based on breed differences in mouth shape and tooth position, are already being produced. Currently available veterinary prescription diets to treat disease are largely aimed at managing the consequences of established disease, and are very broadly applicable. In the future, aetiologically specific diets may become available, although further epidemiological studies are urgently needed to look more closely at the role of diet in disease. As with humans, it is likely that the way we feed our cats as kittens and young adults may well have an impact on the risk of developing degenerative and other diseases in old age.

7. REFERENCES

Adkins, Y., Zicker, S.C., Lepine, A. and Lonnerdal, B. (1997) Changes in nutrient and protein composition of cat milk during lactation. *American J. Veterinary Research* **58**, 370-375.

Anderson, P.A., Baker, D.H., and Corbin, J.E. (1979) Lysine and arginine requirements of the domestic cat. *J. Nutrition* **109**, 1368-1372.

Barnett, K.G. and Burger, I.H. (1980) Taurine deficiency retinopathy in the cat. *J. Small Animal Practice* **21**, 521-526.

Bateson, P. (2000) Behavioural development in the cat. In Turner, D.C. and Bateson, P. (eds.). *The Domestic Cat: the biology of its behaviour*, 2nd edn., Cambridge University Press, Cambridge, pp. 9-22.

Bradshaw, J. (1992) Behavioural Biology. In Thorne, C. (ed.) *The Waltham Book of Dog and Cat Behaviour*, Pergamon Press, Oxford, pp. 31-52.

Bradshaw, J. and Thorne, C. (1992) Feeding Behaviour. In Thorne, C. (ed.). *The Waltham Book of Dog and Cat Behaviour*, Pergamon Press, Oxford, pp. 115-129.

Casal, M.L., Jezyk, P.F. and Giger, U. (1996) Transfer of colostral antibodies from queens to their kittens. *American J. Veterinary Research* **57**, 1653-1658.

Case, L.P., Carey, D.P., Hirakawa, D.A. and Daristotle, L. (2000) *Canine and Feline Nutrition* (2nd ed.). Mosby Inc., Missouri.

Crossley, D.A. (1991) Survey of feline dental problems in a small animal practice in NW England. *British Veterinary Dental Association Journal* **2**, 2-6.

Davidson, M.G. (1992) Thiamin deficiency in a colony of cats. *Veterinary Record* **130**, 94-97.

DiBartola, S.P., Buffington, C.A., Chew, D.J., McLoughlin, M.A. and Sparks, R.A. (1993) Development of chronic renal disease in cats fed a commercial diet. *J. Veterinary Medical Association* **202**, 744-751.

Dickinson, C.D. and Scott, P.P. (1956) Protein requirements for growth of weanling kittens and young cats maintained on a mixed diet. *British J. Nutrition* **10**, 311-316.

Donovan, S.M. and Odle, J. (1994) Growth factors in milk as mediators of infant development. *Annual Review of Nutrition* **14**, 147-167.

Fisher, E.W. (1982) Neonatal diseases of dogs and cats. *British Veterinary Journal* **138**, 277-284.

Fitzgerald, B.M. and Turner, D.C. (2000) Hunting behaviour of domestic cats and their impact on prey populations. In Turner, D.C. and Bateson, P. (eds.). *The Domestic Cat: the biology of its behaviour*, 2nd edn., Cambridge University Press, Cambridge, pp. 151-175.

Gerber, H., Peter, H., Ferguson, D.C. and Peterson, M.E. (1994) Etiopathology of feline toxic nodular goiter. *Veterinary Clinics of North America: Small Animal Practice* **24**, 541-565.

Hand, M.S., Thatcher, C.D., Remillard, R.L. & Roudebush, P. (2000) *Small Animal Clinical Nutrition*. 4th edn., Walsworth Publishing Company, Missouri.

Harper, E.J. (1998) Changing perspectives on aging and energy requirements, aging and digestive function in humans, dogs and cats. *J. Nutrition* **128**, 2623S-2635S.

Hickman, M.A., Rogers, Q.R. and Morris, J.G. (1990) Effect of processing on fate of dietary [14C]taurine in cats. *J. Nutrition* **120**, 995-1000.

Hill, W.F. (1978) Effects of mere exposure on preferences in non-human mammals. *Psychological Bulletin* **85**, 1177-1198.

Houpt, K.A. and Beaver, B. (1981) Behavioural problems of geriatric dogs and cats. *Veterinary Clinics of North America: Small Animal Practice* **11**, 643-652.

Kelly, N. and Wills J. (1996) *BSAVA Manual of Companion Animal Nutrition and Feeding*. British Small Animal Veterinary Association, Cheltenham, UK.

Kirk, S.J. and Barbul, A. (1990) Role of arginine in trauma, sepsis and immunity. *J. Parenteral and Enteral Nutrition* **14**, 226S-229S.

Kitamura, T., Yasuda, J., Hashimoto, A. (1999) Insulin response to intravenous arginine in nonobese healthy cats. *J. Veterinary Internal Medicine* **13**, 549-556.

Klimberg, V.V. and Souba, W.W. (1990) The importance of intestinal glutamine metabolism in maintaining a healthy gastrointenstinal tract and supporting the body's response to injury and illness. *Surgery Annuals* **22**, 61-76.

Lepine, A.J. (1998) Nutrition of the neonatal canine and feline. In Reinhart, G.A. and Carey, D.P. (eds.) *Recent advances in canine and feline nutritional research [volume II]*. Orange Frazer Press, Ohio, pp. 249-255.

Lund, E.M., Bohacek, L.K., Dahlke, L., King, V.L., Kramek, B.A. and Logan, E.I. (1998) Prevalence and risk factors for odontoclastic resorptive lesions in cats. *J. American Veterinary Medical Association* **212**, 392-395.

MacDonald, M.L., Rogers, Q.R. and Morris, J.G. (1984) Role of linoleate as an essential fatty acid for the cat independent of arachidonate synthesis. *J. Nutrition* **113**, 1422-1433.

Mares-Perlman, J.A., Lyle, B.J., Klein, R., Fisher, A.I., Brady, W.E., Van den Langenberg, G.M., Trabulsi, J.N. and Palta, M. (2000) Vitamin supplement use and incident cataracts in a population-based study. *Archive of Ophthalmology* **118**, 1556-1563.

Markham, R.W. and Hodgkins, E.M. (1989) Geriatric nutrition. *Veterinary Clinics of North America: Small Animal Practice* **19**, 165-185.

Morita, T., Awakura, T., Shimada, A., Umemura, T., Nagai, T. and Haruna, A. (1995) Vitamin D toxicosis in cats, natural outbreak and experimental study. *J. Veterinary Medical Science* **57**, 831-837.

Morris, J.G. (1996) Vitamin D synthesis by kittens. *Veterinary Clinical Nutrition* **3**, 88-92.

Morris, J.G. and Rogers, Q.R. (1978) Ammonia intoxication in the near-adult cat as a result of a dietary deficiency of arginine. *Science* **199**, 431-432.

Morris, J.G. and Rogers, Q.R. (1995) Assessment of nutritional adequacy of pet foods through the lifecycle. *J. Nutrition* **124**, S2520-S2534

Mugford, R.A. (1977) External Influences on the Feeding of Carnivores. In Kare, M.R. and Maller, O. (eds.). *The Chemical Senses and Nutrition*, Academic Press, New York, pp. 3-14.

Mugford, R.A. and Thorne, C. (1980) Comparative Studies of Meal Patterns in Pet and Laboratory Housed Dogs and Cats. In Anderson, R.S. (ed.). *Nutrition of the Dog and Cat*, Pergamon Press, Oxford, pp. 25-50.

National Research Council (2003) *Nutrient Requirements of Cats and Dogs*. The National Academies Press, Washington.

Olovson, S.G. (1986) Diet and breeding performance in cats. *Laboratory Animals* **20**, 221-230.

Piechota, T.R., Rogers, Q.R. and Morris, J.G. (1995) Nitrogen requirements of cats during gestation and lactation. *Nutrition Research* **15**, 1535-1546.

Pion, P.D., Kittleson, M.D., Rogers, Q.R. and Morris, J.G. (1987) Myocardial failure in cats associated with low plasma taurine, a reversible cardiomyopathy. *Science* **237**, 764-768.

Raila, J., Mathews, U. and Schweigert, F.J. (2001) Plasma transport and tissue distribution of beta-carotene, vitamin A and retinol-binding protein in domestic cats. *Comparative Biochemistry and Physiology Part A: Molecular Integrated Physiology* **130**, 849-856.

Remillard, R.L., Pickett, J.P., Thatcher, C.D. and Davenport, D.J. (1993) Comparison of kittens fed queen's milk with those fed milk replacers. *American J. Veterinary Research* **54**, 901-907.

Robinson, I. (1992) Behavioural development of the cat. In Thorne, C. (ed.). *The Waltham Book of Dog and Cat Behaviour*, Pergamon Press, Oxford, pp. 53-64.

Robinson, I. (1995) Associations between Man and Animals. In Robinson, I. (ed.). *The Waltham Book of Human-Animal Interaction*, Pergamon Press, Oxford, pp. 1-6.

Scheppach, W. (1994) Effects of short chain fatty acids on gut morphology and function. *Gut* **35**,S35-38

Seawright, A.A. and English, P.B. (1967) Hypervitaminosis A and deforming cervical spondylosis of the cat. *J. Comparative Pathology* **77**, 29-39.

Serpell, J.A. (2000) Domestication and history of the cat. In Turner, D.C. and Bateson, P. (eds.). *The Domestic Cat: the biology of its behaviour*, 2nd edn., Cambridge University Press, Cambridge, pp. 179-192.

Sparkes, A.H., Papasouliotis, K., Sunvold, G., Werrett, G., Clarke, C., Jones, M., Gruffydd-Jones, T.J. and Reinhart, G. (1998) Bacterial flora in the duodenum of healthy cats, and effect of dietary supplementation with fructo-oligosaccharides. *American J. Veterinary Research* 1998 **59**,431-435.

Studer, E. and Stapley, R.B. (1973) The role of dry foods in maintaining healthy teeth and gums in the cat. *Veterinary Medicine – Small Animal Clinician* **68**, 1124-1126.

Sturman, J.A., Gargano, A.D., Messing, J.M. and Imaki, H. (1986) Feline maternal taurine deficiency, effect on mother and offspring. *J. Nutrition* **116**, 655-667.

Taylor, E.J., Adams, C. and Neville, R. (1995) Some nutritional aspects of ageing in dogs and cats. *Proceedings of the Nutrition Society* **54**, 645-656.

Thorne, C. (1982) Feeding behaviour in the cat – recent advances. *J. Small Animal Practice* **23**, 555-562.

Thorne, C. (1992) Evolution and Domestication. In Thorne, C. (ed.). *The Waltham Book of Dog and Cat Behaviour*, Pergamon Press, Oxford, pp. 1-30.

Tidholm, A., Karlsson, I. and Wallius, B. (1996) Feline pansteatitis, a report of five cases. *Acta Veterinaria Scandinavia* **37**, 213-217.

Tomsa, K., Glaus, T., Hauser, B., Fluckiger, M., Arnold, P., Wess, G. and Reusch, C. (1999) Nutritional secondary hyperparathyroidism in six cats. *J. Small Animal Practice* **40**, 533-539.

Watson, A.D. (1994) Diet and periodontal disease in dogs and cats. *Australian Veterinary J.* **71**, 313-318.

Zetner, K. and Steurer, I. (1992) The influence of dry food on the development of feline neck lesions. *J. Veterinary Dentistry* **9**, 4-6.

Chapter 10

BREEDING AND WELFARE

Andreas Steiger
Division of Animal Housing and Welfare, Institute of Animal Genetics, Nutrition and Housing, Vetsuisse Faculty of the University of Bern, CH 3001 Bern, Switzerland

Abstract: The fur, the skeleton, the sense organs and behaviour define the main characteristics of pedigree cat breeds. Breeding for extreme breed types may be associated with adverse effects on the welfare of the queen and her offspring. The aim of breeding should be that both parents and offspring, being in general good health, are capable of their normal species-specific behaviour. According to broad principles of animal protection, considerable deviations in breed morphology, physiology and behaviour, causing pain and suffering, should not be tolerated. An overview of various breeds of cats is presented, together with the major problems associated with these breeds and measures to prevent or treat them. In 1995, within the Council of Europe, an expert committee elaborated a "Resolution on the Breeding of Pet Animals". It includes a detailed description of extreme cat breeds, and encourages breeding associations, breeders, judges and owners to reconsider breeding standards, to interpret them in a responsible way and to raise public awareness to breed-associated problems. The aim of regulations and standards should be to reduce and prevent abuse and suffering in cats caused by the adverse effects of extreme breed characteristics on welfare.

1. INTRODUCTION

Although historically cats were associated with human settlements relatively early, for thousands of years there was little interest in domesticating them. Their reproduction occurred independent of man, which is the main reason why, in contrast to other domestic animals, no special pedigree cat breeds were created for a long time. Even today, the majority of domestic cats do not belong to a particular breed (Christoph 1977; Wegner 1995). The first breeds were rare, and proper breeding of pedigree cats did not occur until the second half of the 19th century (Schwangart 1932; Vella *et al.* 2002, with a chronology of cat breeds). An initial provisional classification of breeds was made in England in 1878 (Herre & Röhrs 1990). Today, a growing number of breeds and colour types exist. Aspects of domestication of cats are mentioned in Chapters 1 and 3 and are treated in detail by Herre and Röhrs (1990), Benecke (1994) and Vella *et al.* (2002). The main characteristics of pedigree breeds are defined by the fur (colour and colour patterns, length and type of hair), the skeleton (legs, tail, head, body), the sense organs (ears, eyes) and the behaviour (for example, levels of activity) (Figure 1). Selecting for extreme breed types may be associated with adverse effects on health and welfare. This chapter deals with such effects, with measures to avoid them and with general principles for responsible breeding.

Figure 1. The main characteristics of pedigree breeds are defined by the fur, the skeleton, the sense organs and the behaviour; this is a tabby-point Siamese. (Courtesy of T. Hearn).

2. EFFECTS OF BREEDING ON WELFARE

For thousands of years man has caused animals to change by domestication (farm animals, horses, companion animals); breeding and rearing wild animals for special purposes, such as the production of meat, milk, eggs and fur; using animals in performance sports; selecting for special appearance or behaviour; using laboratory animals and developing models for disease research. More recently, farm and laboratory animals have also been altered by genetic engineering techniques (Monastersky & Robl 1995; Murray *et al.* 1999).

Animal welfarists and some other scientists have criticized several pedigree breeds of farm animals and companion animals. They are considered to be defective breeds (in German "Qualzuchten"), since they have extreme morphological, physiological or behavioural characteristics. It is claimed or assumed that these characteristics lead to unnecessary suffering and to unacceptable restrictions in the life of these animals. This applies to certain breeds of cattle, swine, horses, dogs, cats, rabbits, pet rodents, poultry, turkeys, ducks, pigeons, pet birds, reptiles, amphibians and ornamental fish (Wegner 1993; Bechtel 1995; Peyer 1997; Wegner 1997; Bartels & Wegner 1998; Not 1998; Peyer & Steiger 1998; Stucki 1998; McGreevy & Nicholas 1999; Expert Group Germany 2000; Bartels 2002). The most active discussion on adverse breeding effects in companion animals has been in the German-speaking community, but discussion is also evident in the English-speaking community where the term "Animal Illfare" has been used (Ott 1996). Reviews on extreme breed types in domestic animals are presented by Bartels and Wegner (1998), in companion animals by a report of an Expert Group in Germany (2000), in dogs by Peyer (1997, a literature review in German), in cats by Wegner (1995) and Stucki (1998, a literature review in German) and in exhibition poultry and cage birds by Bartels (2002). A detailed introduction to cat genetics for cat breeders can be found in Vella *et al.* (2002) in the book "Robinson's genetics for cat breeders and veterinarians", with chapters on reproduction, genetics and heredity, breeding systems and practices.

3. GENERAL PRINCIPLES OF RESPONSIBLE BREEDING

In the "European Convention for the Protection of Pet Animals" of 13th November 1987, the Council of Europe presented the basic principles of correct breeding of pet animals (Council of Europe 1987). It states that breeders are responsible for considering the anatomical, physiological and behavioural characteristics that are likely to put the health and welfare of either the offspring or the female parent at risk. Accordingly, and considering general principles of animal welfare, some basic rules for responsible breeding can be established:

- to breed animals, both parents and offspring, that are generally in good health and capable of their normal species-specific behaviour,
- to avoid unnecessary pain, suffering and damage of the animals, due to breeding as direct inherited breed effects or indirectly as the effect of special housing conditions,
- to avoid considerable deviations in morphology (e.g. organs, limbs), in physiology (e.g. sense organs) and in behaviour of the breed type, compared to other breed types of the same species (e.g. lack of tail, deafness), and to avoid the impairment of basic functions of the organs, of parts of the body and of behaviour,
- to avoid inherited diseases (no increased morbidity) and to avoid increased mortality (including stillbirths, lethal factors), compared to other breed types of the same species, especially by avoidance of inbreeding,
- in the case of animals used for reproduction: to assure the possibility of species-specific reproduction, including normal mating behaviour, normal parturition, and normal rearing of the offspring (e.g. avoidance of routine Caesarean sections, promoting good maternal instincts),
- to assure the possibility of species-specific behaviour of locomotion, including climbing and jumping (by avoiding, in particular, short legs) and of food intake (by avoiding particular anatomical anomalies of mouth and teeth),
- to avoid considerable impairment of behaviour and to avoid special requirements for housing conditions due to the breed type (particularly naked animals, animals with difficulties of body temperature regulation),
- to respect the biological limits of adaptation of the animals to the effects of breeding.

These general principles should be the basis for practical breeding, breed standards and for legislation.

4. ETHICAL ASPECTS AND BREEDING REGULATIONS

General ethical principles, as well as national regulations on animal protection, require the avoidance of unnecessary suffering, pain and damage of animals in general, in particular with regard to the housing and treatment of animals. These principles may be applied to breeding methods, too. Some national regulations, for example in Germany (in paragraph 11b of the German law on animal protection, 1986), have included special provisions concerning breeding. Breeding of animals by traditional methods or by genetic engineering is prohibited, if parts of the body or organs are lacking or may be altered in such a way that pain, suffering, damage, behavioural disorders or increased aggressiveness occur in the offspring or the parents, or if housing of the animals is only possible under conditions leading to pain, suffering or damage.

The "European Convention for the Protection of Pet Animals" of the Council of Europe, based in Strasbourg, states the following (Council of Europe 1987, article 5):

Any person who selects a pet animal for breeding shall be responsible for having regard to the anatomical, physiological and behavioural characteristics, which are likely to put at risk the health and welfare of either the offspring or the female parent.

The explanatory report of this Convention comments on this rule as follows (Council of Europe 1987):

Article 5 lays down the principle that, in the breeding of pet animals, care should be taken by those responsible for the breeding to ensure that the physical and mental health of the offspring and female parent are not put at risk. In the selection of specimens for breeding, care should be taken to avoid the transmission of behavioural patterns such as abnormal aggressive tendencies and hereditary defects: for example progressive retinal atrophy (leading to blindness), oversized foetal heads (preventing normal birth), and other physical characteristics required by certain breed standards which predispose to clinical conditions such as entropion and soft-plate deformities.

On the basis of this Convention and its general rules, an expert committee in the Council of Europe elaborated and adopted a "Resolution on the Breeding of Pet Animals" in 1995 (Council of Europe 1995a). This paper provides recommendations for the application and interpretation of the general rules in article 5 of the Convention. It includes detailed descriptions of extreme breed types of dogs and cats, with examples of various breeds, and it asks and encourages breeding associations, including breeders, judges and owners, to reconsider breeding standards; to select the animals not only taking into account aesthetic criteria; to ensure, by good information and education, the interpretation of breeding standards in a responsible way and to raise public awareness to the breed problems. The resolution, which is not sufficiently widely known and applied, is presented in Table 1 in its full length, with provisions concerning only cats or cats in particular in *italic*.

Table 1. Council of Europe, Resolution on the Breeding of Pet Animals (adopted on 10th March 1995 by the Multilateral Consultation of Parties to the European Convention for the Protection of Pet Animals)

The Parties of the European Convention for the Protection of Pet Animals, by virtue of the terms of reference laid down in Article 15;

- Recognising that these terms of reference imply the monitoring of the implementation of the provisions of the Convention and the development of common and co-ordinated programmes in the field of pet animal welfare;
- Anxious to encourage full respect of the provisions of the Convention;
- Recalling that Article 5 of the Convention provides for a selection of pet animal for breeding which takes account of the anatomical, physiological and behavioural characteristics which are likely to put at risk the health and welfare of either the offspring or the female parent;
- Aware that problems are encountered with the implementation of these provisions, in particular with the development of extreme characteristics detrimental to the health and welfare of the animals;
- Convinced that these problems are related for a large part to the way breeding standards are formulated and interpreted;
- Considering therefore that a revision of these breeding standards is necessary in order to fulfil the requirements of Article 5 of the Convention;

Agreed:
1. to encourage breeding associations, *in particular cat and dog breeding associations*:
- to reconsider breeding standards in order, if appropriate, to amend those which can cause potential welfare problems, in particular in the light of the recommendations presented in the Appendix;
- to reconsider the standards and to select the animals taking into account not only aesthetic criteria but also behavioural characteristics (for instance with regard to problems of aggressiveness) and abilities;
- to ensure, by good information and education of breeders and judges, that breeding standards are interpreted in such a way as to counteract the development of extreme characteristics ("hypertype") which can cause welfare problems;
- to raise public awareness to the problems related to some physical and behavioural characteristics of the animals;
2. if these measures are not sufficient, to consider the possibility of prohibiting the breeding and for phasing out the exhibition and the selling of certain types or breeds when characteristics of these animals correspond to harmful defects such as those presented in the Appendix.

Appendix
- The Parties are convinced that in the breeding of several breeds or types of pet animals, mammals and birds, insufficient account is taken of anatomical, physiological and behavioural characteristics which are likely to put at risk the animals' health and welfare.
- However, the Parties considered that *problems connected with the breeding of cats* and dogs should be addressed as a priority.
- The Parties strongly encourage cat and dog breeding associations to revise their breeding policies in the light of Article 5 of the Convention taking account in particular of the following guidelines:

Guidelines for the revision of breeding policies:
(The breeds mentioned in brackets are only examples in which these problems may occur) –
- set maximum and minimum values for height or weight of very large or small dogs, respectively, to avoid skeleton and joint disorders (e.g. dysplasia of hip joints or elbows, fractures, luxation of elbow or patella, persistent fontanella) and collapse of trachea;
- set maximum values for the proportion between length and height of short-legged dogs (e.g. Bassethound, Dachshund) to avoid disorders of the vertebral column;
- *set limits to the shortness of skull, and in particular the nose, so that breathing difficulties and blockage of lachrymal ducts are avoided, as well as disposition to birth difficulties (e.g. Persian Cats, especially the "extreme type",* Bulldogs, Japan Chin, King Charles Spaniel, Pug, Pekin Palacedog);
- prevent the occurence of:
-- a persistent fontanella (e.g. Chihuahua) to avoid brain damages;
-- abnormal positions of legs (e.g. very steep line of hind legs in Chow Chow, Norwegian Buhund, Swedish Lapphund, Finnish Spitz; bowed legs in Bassethound, Pekin Palacedog, Shi Tzu) to avoid difficulties in movement and joint degeneration;
-- *abnormal positions of teeth (e.g. brachygnathia in* Boxers, Bulldogs, *Persian Cats) to avoid difficulties in feeding and caring for the newborn;*
-- abnormal size and form of eyes or eyelids (e.g. ectropion: Bassethound, Bloodhound, St. Bernard;
-- small deep lying eyes with disposition to entropion: Airedale Terrier, Australian Terrier, Bedlington Terrier, Bullterrier, Bloodhound, Chow Chow, English Toy Terrier, Jagdterrier, Newfoundland, Shar Pei; large, protruding eyes: Boston Terrier, Cavalier King Charles Spaniel, Dandie Dinmont Terrier, Brussels Griffon, Japan Chin, King Charles Spaniel, Pug, Pekin Palacedog, Shi Tzu, Tibet Terrier) to avoid irritation, inflammation and degeneration as well as prolaps of eyes;
-- very long ears (e.g. English Cocker Spaniel, Bassethound, Bloodhound) to avoid disposition to injuries;
-- markedly folded skin (e.g. Bassethound, Bulldog, Bloodhound, Pug, Pekin Palacedog, Shar Pei) to avoid eczemas and in the case of furrows around the eyes irritation and inflammation of eyes;
- avoid or, if it is not possible to eliminate severe defects, discontinued breeding of:
-- animals carrying semi-lethal factors (e.g. Entlebucher Cattledog);
-- *animals carrying recessive defect-genes (e.g. homozygous Scottish Fold Cat: short legs, vertebral column and tail defects)*
-- *hairless dogs and cats (lack of protection against sun and chill, disposition to significant reduction of number of teeth, semi-lethal factor)*
-- *Manx-cat (movement disorder, disposition to vertebral column defects, difficulties in elimination of urine and faeces, semi-lethal factor)*
-- *cats carrying "dominant white" (significant disposition to deafness);*
-- dogs carrying "Merle factor" (significant disposition to deafness and eye disorders, e.g.: Blue merle Collie, Merle Sheltie, Merle Corgie, Merle Bobtail, Tigerdogge, Tigerteckel).

As a consequence of this "Resolution on the Breeding of Pet Animals" the parties involved in its elaboration, including four international breeding associations, also agreed on a "Declaration of Intent" at the Council of Europe in 1995. It was adopted at the same Multilateral Consultation of Parties to the European Convention for the Protection of Pet Animals on 10th March 1995. The declaration, which is also not sufficiently widely known and applied, is presented in Table 2 (Council of Europe 1995b).

Table 2. Declaration of Intent (adopted on 10th March 1995 at the Multilateral Consultation of Parties to the European Convention for the Protection of Pet Animals)

The Parties to the European Convention for the Protection of Pet Animals and the Fédération Cynologique Internationale, the *Fédération Internationale Féline, the Governing Council of the Cat Fancy and the World Cat Federation* agreed on the need to *improve breeding and breeding standards of cats and* dogs in accordance with the principles set out in the Convention. In particular, they agreed: - to contribute to the improvement of breeding standards, in particular with regard to surgical operations for aesthetic purposes, taking in account the welfare of the animals; - to promote the respect of these standards by the judges and the breeders; - to facilitate the appropriate and continuing training of judges and breeders; - to take necessary measures to control the breeding of animals with genetic or phenotypic characteristics harmful to the welfare of the animals in order to prevent suffering of such animals; - to develop information to the public in order to achieve responsible ownership in accordance with the provisions of the Convention.

Both the "Resolution on the Breeding of Pet Animals" of the Council of Europe and the "Declaration of Intent" emphasize the importance of contributions from several partners, such as breeding associations, breeders, judges and owners, the state and authorities, to improve aspects of animal welfare in breeding.

The Federation of Veterinarians of Europe (FVE) has also issued a resolution on "Breeding and Animal Welfare" which "urges its member countries and the European Commission to consider the introduction of measures designed to safeguard the welfare of animals with respect to the risks inherent in selective breeding programmes, while preserving the unique characteristics and genetic advantages of European breeds" (Federation of Veterinarians of Europe 1999). The resolution continues: "Selective breeding programmes may cause animal welfare problems. It may become difficult or impossible for natural copulation or parturition to occur; offspring produced by selective breeding for certain specific, characteristics may be unable to express their natural behaviour; or they may be predisposed to hereditary, congenital, metabolic or infectious disease, disability or early death. The introduction and continuation of such selective breeding programmes may make it impossible for the breed to be maintained by natural means (...) The FVE believes that it is the function of the veterinary profession not only to treat sick and injured animals, but to promote and safeguard animal health and welfare. Its members believe that selective breeding of animals should not be used to introduce a welfare deficit into a species or breed, or to impair the ability of a breed or individual to express its natural behaviour throughout its natural lifespan. Furthermore, where selective breeding has already resulted in welfare disadvantages being introduced into any species or breed, the FVE urges veterinarians not only to treat individual animals humanely, but also to bring to the attention of the breeding organizations and the competent authorities in their countries the need for action to alleviate the welfare problems caused by selective breeding".

5. WELFARE OF PEDIGREE BREEDS

Adverse effects on health and welfare are known to be associated with several cat breeds. Table 3 gives an overview of various breeds of cats with such welfare problems, divided into the main categories of skin, skeleton (including head) and sense organs, and includes the major problems of health and welfare, examples of affected breeds and measures proposed to avoid these problems. A glossary of genetic terms is presented in Table 4. An example of a brachycephalic breed, the Persian, is shown in Figure 2.

Figure 2. This Colourpoint Persian kitten has a relatively big, round head and short, broad nose. The upper part of the nasal plate is higher than the level of the lower eye-lids. This type of conformation can cause welfare problems (see Table 3). (Courtesy of Thomas Bartels).

Table 3. Overview of various cat breeds with welfare problems, divided into the main categories of breed characteristics, with the major symptoms and welfare problems, genotype (where known), examples of affected breeds, and proposed measures to avoid the problems and improve welfare.
References: (1) Council of Europe 1995a, (2) Bartels & Wegner 1997, (3) Keller 1997, (4) Stucki 1998, (5) Expert Group Germany 2000, (6) Vella *et al.* 2002, (7) Wegner 1995.

Breed characteristics (Phenotype)	Symptoms and associated welfare problems	Genetics (Genotype) (see Glossary)	Breeds where problems may occur	Measures to improve welfare
SKIN				
Long hair	Regular care necessary, thermoregulation may be impaired, occurrence of eczemas	Longhair gene (2), homozygous (6)	Persian, Angora	Careful selection of owners, regular care necessary
Short hair (Rex-type)	Reduced growth of hairs (hypoplasia), hairs shorter and thinner than normal, unequal structure, wavy or ruffled fur. Cornish and German-Rex: primary guard hairs missing, awn hair altered, short and curly fur. Devon Rex: hairs fragile, partial hairlessness, whiskers curled or missing. Thermoregulation may be impaired, impaired behaviour (e.g. social communication, hunting).	Autosomal, recessive, Cornish and German Rex (gene r), Devon Rex (gene re)	Cornish Rex, Devon Rex, German Rex and others.	Prohibition of breeding and exhibition for cats with lack of or defective whiskers or with partially hairless skin (4), prohibition of breeding of cats with lack of whiskers, change of standards to avoid animals with shortened or curled whiskers (5)
Hairlessness	Disturbed growth of hair (hypotrichosis congenita), few thin hairs on some parts of body, in particular nose and legs, whiskers may be absent, skin with thickening, many folds, tendency to higher production of sebum. Lack of protection against sun and chill, disposition to significant reduction in number of teeth, semi-lethal factor, impaired thermoregulation, impaired behaviour (social communication, comfort behaviour, orientation, prey-catching).	Autosomal, recessive French Hairless (gene h), Redcar Hairless (gene hd), Canadian Hairless (gene hr, Sphinx) (6)	Sphinx, French Hairless, Redcar Hairless, Canadian Hairless	Avoid and eliminate defects, discontinue breeding, change of breed standards (1), prohibition of breeding and exhibition for all hairless cats (4), prohibition of breeding of cats with lack of whiskers, change of breed standards (5)

Table 3. (continued). Overview of various cat breeds with welfare problems, divided into the main categories of breed characteristics, with the major symptoms and welfare problems, genotype (where known), examples of affected breeds, and proposed measures to avoid the problems and improve welfare.

Breed characteristics (Phenotype)	Symptoms and associated welfare problems	Genetics (Genotype) (see Glossary)	Breeds where problems may occur	Measures to improve welfare
White Fur	Severe disposition to deafness in cats with white fur caused by the dominant white gene (leukism), impaired hearing to total deafness, lack of pigmentation and tapetum lucidum in the eyes, higher incidence of skin cancer (5, 6), impaired behaviour (social communication, impaired attention to environmental factors). Cats with totally white fur associated with an extreme form of spotting (gene S) might also show deafness (3). Animals with white fur determined by genes of the albino-series (gene c) are not likely to suffer from deafness, except when combined with the W-gene (Foreign White).	Autosomal, dominant white (gene W), epistatic (suppresses all other colour genes), pleiotropic (causing white fur in 100%, blue eyes in 70% and deafness in 50% of cases)	White cats, all cat breeds with white type, e.g. European Short Hair, British Short Hair, Norwegian Forest Cat, Maine Coon, Turkish Angora, Persian, Foreign White, Russian White, Van Cat	Avoid mating of two white animals (4). Avoid and eliminate occurrence of defects, discontinue breeding of cats carrying "dominant white" (1), prohibition of breeding and exhibition of cats carrying dominant gene W (4), prohibition of breeding of cats carrying dominant gene W, in unclear cases gene analyses, tattoo or microchip for all examined animals (5), behavioural, ophthalmologic and audiometric examination (electrical reaction audiometry) of cats before use for breeding is very important (5, 6)
SKELETON				
Body size	Extreme body size might cause problems with joint articulations like in giant dogs.		Maine Coon, Norwegian Forest Cat	Careful selection of suitable parents, no over-typing
Brachycephalia, short head	Big, round head with a short broad nose, in extreme cases upper part of nasal plate is higher than the level of the lower eye-lids (peke-face); brachygnathia, abnormal position of teeth, disposition to birth difficulties (2). Short upper jaw (brachygnathia superior), narrow upper respiratory passages (nasal	Polygenic, heterogeneity possible (2)	Persian cats, Exotic Shorthair, Burmese breed	Set limits to shortness of skull and nose, prevent the occurrence of abnormal teeth positions, change of standards (1), change of breed type and of standards, no over-typing, prohibition of breeding and exhibition of animals with defects (4), change of

Table 3. (continued). Overview of various cat breeds with welfare problems, divided into the main categories of breed characteristics, with the major symptoms and welfare problems, genotype (where known), examples of affected breeds, and proposed measures to avoid the problems and improve welfare.

Breed characteristics (Phenotype)	Symptoms and associated welfare problems	Genetics (Genotype) (see Glossary)	Breeds where problems may occur	Measures to improve welfare
	stenosis, alterations of septum nasalis), breathing difficulties, narrow or blocked lachrymal ducts (in particular the extreme peke-face types), entropion of eye-lids, tendency to dystocia, higher rate of stillbirth, impaired behaviour (social communication, comfort behaviour), difficulties in feeding and in caring for the newborn.			standards, elaboration of an index to avoid over-typing, prohibition of breeding of animals outside these limits, clinical examination of animals before breeding (5)
Entropion	Irritation of cornea and conjunctiva	Polygenic, occurring in particular in connection with brachycephalia	Sporadic in many breeds, especially in brachycephalic types	Prohibition of breeding with animals with entropion (5)
Short tail, taillessness	Manx and Cymric: shortening of tail in various degrees up to a dent (incisure or hollow) instead of a tail, hind legs elevated, hopping locomotion. Malformation of vertebrae in various expressions, anatomical and neurological disorders of the pelvic area and spinal cord, including rectum and anus; movement disorders, disposition to vertebral column and other defects, difficulties in elimination of urine and faeces (1), impaired behaviour, extreme painfulness of the pelvic area in tailless animals (5). Japanese Bobtail, Kuril Bobtail: shortened rolled tails, tail-process of 10-15 cm, only	Manx and Cymric: Gene M autosomal, incomplete dominant inheritance, various expressivity with range from wild type with tail (gene mm) to shortened tail in various expressions to lethal (gene MM, early embryonic death) (6, 2) Bobtail: probably recessive or incomplete dominant	Manx, Cymric, Japanese Bobtail, Kuril Bobtail, sporadic in all breeds	Avoid and eliminate occurrence of defects, discontinue breeding (1), prohibition of breeding and exhibition of animals carrying those characteristics and carriers of taillessness, also for short-tailed cats (4) Manx and Cymric: prohibition of breeding, Japanese Bobtail and Kuril Bobtail: recommendation to breed associations for examination of animals for pain sensitivity and for vertebral column defects before breeding, tattoo or microchip for examined animals (5)

Table 3. (continued). Overview of various cat breeds with welfare problems, divided into the main categories of breed characteristics, with the major symptoms and welfare problems, genotype (where known), examples of affected breeds, and proposed measures to avoid the problems and improve welfare.

Breed characteristics (Phenotype)	Symptoms and associated welfare problems	Genetics (Genotype) (see Glossary)	Breeds where problems may occur	Measures to improve welfare
	movable at the base, vertebrae merged.	heredity without lethal effect		
Short legs	Dwarfism, reduced growth of limb bones, chondrodysplasia, chondrodystrophy. Impaired locomotion (especially jumping and climbing), occasionally tendency to defects of intervertebral disks (7).	Probably autosomal, incomplete dominant inheritance, eventually polygenic (5), polyfactorial (6)	Munchkin (dachshund cat)	Prohibition of breeding and exhibition (4), renunciation of breeding, further observation of the population for defects, tattoo or microchip for examined animals (5)
Polydactylia	Supernumerary toes, mainly on the front paws, hind legs can also be altered, semilethal factor (6).	Gene Pd, autosomal, dominant with variable expression, semi-lethal factor (2)	Sporadic in all breeds, frequently in Maine Coon ("Super-scratcher")	Prohibition of breeding of animals carrying characteristic (5)
Microbrachia	Shortness and deformity of fore limbs, "kangoroo legs", impaired locomotion.	Monogenic, mostly autosomal, recessive (6)	"Twisty Cat" "Kangoroo Cat"	Prohibition of breeding of animals carrying characteristic (5)

SENSE ORGANS

Fold ears	Pinna of the ears tipped forwards, severe painful defects of bones and cartilage, chondrodystrophy, osteodystrophy, dysplasia of epiphysis in homozygote and sometimes in heterozygote animals (5), impaired behaviour (social communication). Recessive defective gene, homozygote: short legs, vertebral column and tail defects (1).	Incomplete dominant (gene Fd), incomplete penetrance: homozygous (gene FdFd) always with fold ears and defects of bones and cartilage, heterozygous (gene Fdfd) mostly with	Scottish Fold, Highland Fold, Poodle cat, sporadic in other breeds	No breeding of homozygote cats, checking of all heterozygote cats for damage of bone and cartilage (may be present without folded ears), exclusion of such animals from breeding. Avoid and eliminate occurrence of defects, discontinue breeding (1), prohibition of breeding and exhibition (4), prohibition of breeding of animals with fold ears determined by the gene

Table 3. (continued). Overview of various cat breeds with welfare problems, divided into the main categories of breed characteristics, with the major symptoms and welfare problems, genotype (where known), examples of affected breeds, and proposed measures to avoid the problems and improve welfare.

Breed characteristics (Phenotype)	Symptoms and associated welfare problems	Genetics (Genotype) (see Glossary)	Breeds where problems may occur	Measures to improve welfare
		fold ears and defects of bones and cartilage (5), normal ears also possible (6)		Fd (5)
Curled ears	Pinna of the ears tipped backwards, other defects not known.	dominant (gene Cu)	American Curl	Careful assessment of future use for breeding (4)
Necrosis of cornea	Focal necrosis of cornea, "mummification" of cornea, pain.		Siamese, Persian, Colourpoint	No breeding, further examination (4)
FURTHER DEFECTS AND DISEASES	Monogenic: congenital lenticular cataracts, chediak-higashi-syndrome, (cyclic neutropaenia), gangliosidosis, haemophilia, cerebral hernia (meningo-encephalocele), congenital diaphragmatic hernia, deformed tails, mucopolysaccharidosis, muscle dystrophy, polycystic kidney disease, progressive retinal atrophy, dental anomalies (hypo- and hyperdontia) (5). Oligo- or polygenic: Brachygnathia superior and inferior, face malformations (also monogenic), hip dysplasia, key-gaskell-syndrome (dysautonomia of neural system), osteogenesis imperfecta, patella luxation.	Monogenic, mostly autosomal, recessive only detectable in animals homozygous for the defective gene (except haemophilia), autosomal dominant in polycystic kidney disease in Persians and Exotic Shorthairs, Oligo- or polygenic	These defects may occur sporadically in families of many breeds, polycystic kidney disease is common in Persians and Exotic Shorthairs	Prohibition of breeding with animals showing or carrying the characteristic, avoiding breeding with relatives, animals used for breeding to be tested for freedom from defects (5) Prohibition of breeding with animals carrying characteristic, or other breeding measures (details see 5)

Table 4. Glossary of genetic terms (according to Vella *et al.* 2002)

autosomal	gene present on the autosomes, as opposed to gene on the sex chromosomes
dominant	expression of a gene completely over-riding that of another gene at the same locus
epistasis	when the expression of an allele masks the effects of one or more alleles at different loci, not to be confused with dominance
expression	variation of the phenotype of a gene
gene	basic unit of heredity, as carried by the chromosomes like 'beads on a string'
heredity	transmission of inherited variation from parent to offspring
heterogeneity	the same phenotype is caused by different genes
heterozygous	two dissimilar alleles at a locus, e.g. Aa
homozygous	possessing a pair of identical alleles at a given locus, e.g. AA or aa
inbreeding	mating closely related individuals, the two most common are brother sister or parent offspring
incomplete dominance	when the heterozygote Aa has a phenotype which is intermediate in expression to those of the two homozygotes AA and aa
inherited	that which is transmitted from parent to offspring
lethal gene	gene which causes death of the recipient
monogenic	character which is determined by a single major gene
oligogenic	character determined by a few genes
penetrance	proportion of individuals that express a particular gene
pleiotropism	when a gene has several apparently unrelated effects on the phenotype, usually these can be traced to a common genetic cause
polygenic	character determined by several genes, usually with reference to minor genes or polygenes
recessive	allele whose expression is over ridden by another at the same locus

6. IMPROVING WELFARE AND BREEDING IN THE FUTURE

In many discussions and publications, criticism of extreme breeds types often involves generalizations and the prohibition of the breed, for example particular breeds with unusual skin, fur or morphology. However, every breed includes variations, with breed characteristics expressed to a varying degree, and lines exist within the breed. It is therefore preferable to limit the descriptions of those breeds to be avoided, rejected or even formally prohibited by legislation, not on the breed itself but on the more precise description of the biological criteria (morphological, physiological and behavioural characteristics) that are acceptable or unacceptable, based on animal welfare considerations. Using these biological criteria specific breeds, their variations, hybrids and future new breeds can be assessed.

Regulations associated with legislation have to be quite general in nature, as in article 5 of the European Convention for the Protection of Pet Animals (Council of Europe 1987), in order to include all possible cases in practice. More detailed regulations should be reserved for guidelines and recommendations, whether included in legislation or not (for example, aimed at breeder associations). The "Resolution on the Breeding of Pet Animals" of the expert committee of the Council of Europe (Council of Europe 1995a) is an example of detailed regulations at the level of guidelines, although even more detailed rules may be required for practical application.

Such detailed regulations should include the following points:

- Criteria (detailed biological characteristics) for the assessment of certain breed types, with description of the limits of an acceptable and an unacceptable breed type; with these criteria breed types and hybrids would be included, and general principles as well as the practical assessment of individuals would be determined.
- Prohibition of breeding of certain breed types (there may be a problem with the importation of animals), with reasonable transition periods for the application of this breeding ban (time for introduction of rules, for information and education measures, for changing breed standards), and with exceptions for animals already alive at the beginning of a ban.
- Prohibition of the exhibition, trade and import of certain breed types (there may be difficulties with the application and control of regulations).
- Examination of some general ethical problems associated with undesirable breeds, which do not do necessarily involve suffering (e.g. lethal factors).
- Instead of prohibiting only breeding, examination of the possibilities of a general prohibition of housing of certain extreme breed types (this would include indirectly prohibitions both of breeding, exhibition and import), with transition periods and exceptions for animals already alive at the beginning of a ban.
- Examination of some adverse breed effects that may be compensated for by special care, special housing conditions (e.g. adequate temperature conditions for nude animals) and special feeding regimes.
- Obligation to undertake adequate measures to inform and educate pet owners.

The aims of these regulations should be to prevent the abuse and unnecessary suffering of animals, without prohibiting the responsible use of animals as companion animals and, if scientifically, ethically and legally justified, in research.

7. CONCLUSIONS

Within Europe legislation now exists, aimed at safeguarding the welfare of pet animals used for breeding purposes and their offspring. In many countries, however, specific regulations on animal protection with regard to breeding are lacking or are only in preparation. There is great public and media interest in animal welfare issues. The veterinary community, breeding associations, breeders, owners, judges, organizers of cat exhibitions, committees responsible for breeding standards, and all those concerned with animal welfare should re-examine their ethical principles in line with the evolving modern attitude towards animals, in this instance cats, as sentient beings, and acknowledge the increasing responsibility of society for their welfare.

8. REFERENCES

Bartels, T. and Wegner, W. (1998) *Fehlentwicklungen in der Haustierzucht.* Enke Verlag, Stuttgart, Germany.
Bartels, T. (2002) Hereditary effects and predispositions in exhibition poultry and cage birds – Erbschäden und Dispositionen bei Rassegeflügel und Ziervögeln, Habilitationsschrift Vetsuisse Faculty of University of Bern, CH 3001 Bern.
Bechtel, H.B. (1995) *Reptile and amphibian variants - colors, patterns and scales.* Krieger Publishing Co., Malabar, Florida.
Benecke, N. (1994) *Der Mensch und seine Haustiere.* Theiss Verlag, Stuttgart, Germany.
Christoph H.J. (1977) *Klinik der Katzenkrankheiten.* Gustav Fischer, Stuttgart, Germany.
Council of Europe (1987) European Convention for the protection of pet animals, 13th November 1987 (ETS 125), Council of Europe, F 67075 Strasbourg-Cedex.
Council of Europe (1995a) Resolution on the breeding of pet animals, Multilateral Consultation of parties to the European Convention for the protection of pet animals (ETS 123), March 1995 in Strasbourg, Document CONS 125(95)29, Council of Europe, F 67075 Strasbourg-Cedex.
Council of Europe (1995b) Declaration of intent, Multilateral Consultation of parties to the European Convention for the protection of pet animals (ETS 123), March 1995 in Strasbourg, Document CONS 125(95)29, Council of Europe, F 67075 Strasbourg-Cedex.
Expert Group Germany, Sachverständigengruppe BML (2000) Gutachten zur Auslegung von § 11b des Tierschutzgesetzes, Bundesministerium für Ernährung, Landwirtschaft und Forsten BML, Rochusstr. 1, D 53107 Bonn. Website at www.bml.de
Federation of Veterinarians of Europe, FVE (1999) Resolution "Breeding and Animal Welfare", FVE/99/010, Website at www.fve.org
Herre W. and Röhrs M. (1990) *Haustiere – zoologisch gesehen.* Gustav Fischer, Stuttgart, Germany.

Keller, P. (1997) Untersuchungen zur Entwicklung der frühen akustisch evozierten Potentiale (FAEP) bei der Katze für den Einsatz in der Grundlagenforschung und zur klinischen Anwendung, Dissertation med. vet. University of Hannover.

McGreevy, P.D. and Nicholas, F.W. (1999) Some practical solutions to welfare problems in dog breeding. *Animal Welfare* **8**, 329-341.

Monastersky, G.M. and Robl, J.M. (1995) *Strategies in transgenic animal science.* American Society for Microbiology Press , Washington.

Murray, J.D., Anderson, G.B., Oberbauer, A.M. and McGloughlin, M.M. (1999) *Transgenic animals in agriculture.* CAB International, Wallingford, Oxon, UK.

Not, I. (1998) Beurteilung verschiedener Zuchtlinien von Ziervögeln, Kleinnagern, Zierfischen und Reptilien in tierschützerischer Hinsicht, Diss. med. vet. University of Zürich (Abteilung für Zoo-, Heim- und Wildtiere, Winterthurerstr. 268, CH 8057 Zürich; available also at Institut für Genetik, Ernährung und Haltung von Haustieren, Bremgartenstr. 109a, CH 3001 Bern).

Ott, R. (1996) Animal selection and breeding techniques that create diseased populations and compromise welfare. *J. American Veterinary Medical Association* **208**, 1969-1974.

Peyer, N. (1997) Die Beurteilung zuchtbedingter Defekte bei Rassehunden in tierschützerischer Hinsicht, Diss. med. vet. University of Bern (Institut für Genetik, Ernährung und Haltung von Haustieren, Bremgartenstr. 109a, CH 3001 Bern).

Peyer, N. and Steiger, A. (1998) Die Beurteilung zuchtbedingter Defekte bei Rassehunden in tierschützerischer Hinsicht, *Schweizer Archiv für Tierheilkunde* **140**, 359-364.

Schwangart F. (1932) *Zur Rassebildung und Rassezüchtung der Hauskatze.* Gustav Fischer, Stuttgart, Germany.

Stucki, F. (1998) Die Beurteilung zuchtbedingter Defekte bei Rassegeflügel, Rassetauben, Rassekaninchen und Rassekatzen in tierschützerischer Hinsicht, Diss. med. vet. University of Bern (Institut für Genetik, Ernährung und Haltung von Haustieren, Bremgartenstr. 109a, CH 3001 Bern).

Vella, C.M., Shelton, L.M., McGonagle, J.J. and Stanglein, T.W. (2002) *Robinson's genetics for cat breeders and veterinarians.* Butterworth Heinemann, Oxford.

Wegner, W. (1993) Tierschutzrelevante Missstände in der Kleintierzucht – der § (paragraph) 11b des Tierschutzgesetzes greift nicht. *Tierärztliche Umschau* **48**, 213-222.

Wegner, W. (1995) *Kleine Kynologie* (with appendix on cat breeding p. 353-400). Terra Verlag, Konstanz, Germany.

Wegner, W. (1997) Tierschutzaspekte in der Tierzucht. In Sambraus, H.H. and Steiger, A. (eds.). *Das Buch vom Tierschutz*, Enke Verlag, Stuttgart, Germany, pp. 556-569.

INDEX

ACTH, 27, 41
adoption of cats, 76, 135, 137, 150
adrenocorticotropic hormone. *See* ACTH
affiliative behaviours, 2, 5, 7, 18, 66
African wild cat, 1, 27, 47, 181
aggression, 2, 4, 12, 16, 18, 29, 30, 66, 76, 92, 94, 95, 99, 102, 108, 110, 112, 222
ailurophobia. *See* fear of cats
allergies and asthma, 72, 126, 130
allogrooming, 7, 66. *See also* affiliative behaviours
allorubbing, 9, 15, 19, 66. *See also* affiliative behaviours
allowing cats outdoors. *See* keeping cats indoors
animal abuse and hoarding, 73, 74
animal law, 75
animal rights, 75
animal rights movement, 53
animal sanctuary. *See* shelter
Animal Welfare Bill, 178
anthropomorphism, 54, 74, 124

attachment
  to cats, 58
  to dogs, 57, 58
autonomic activity, 26, 111
behaviour problems, 25, 38, 67, 71, 75, 77, 80, 92, 129, 132, 158, 195, 216
  cause of euthanasia, 93, 120
  incidence, 94, 134
  treatment, 100
behavioural testing, 37
breeding
  cats, 56, 124, 206, 259, 263
  dogs, 56, 263
  effects on welfare, 261, 266
  ethical principles, 262
  future developments, 273
  responsibilities, 262, 274
Burmese, 94
Cat-Approach-Test. *See* behavioural testing
cat-human interactions, 59, 65, 66, 67, 70, 196, 198, 242
catnip, 185
cats in research, 54, 188
Cat-Stress-Score. *See* CSS

cattery, 25, 28, 178
colony, 2, 3, 4, 6, 19, 107, 144
colostrum, 243
Comfort from Companion Animals Scale, 58
companion animal. *See* pet ownership
companion animal bond, 54, 56, 58
competition for resources, 28, 145. *See also* food resources
corticosteroid. *See* cortisol
cortisol, 26, 39, 42, 101, 216
CSS, 32, 42, 191, 192, 193
Declaration of Intent, 265. *See also* breeding
defaecation. *See* elimination behaviour
dermatological conditions, 112
disease and behaviour, 110
domestic environment. *See* home
domestication, 27, 28, 47, 73, 228, 229, 260, 261
dominance, 12, 13, 14, 65, 110, 198
ear tipping, 150
early-age spaying and neutering, 124
elimination behaviour, 2, 17, 38, 76, 95, 103
emotions. *See* feelings
enrichment. *See also* housing requirements
auditory, 186
environmental, 181, 185, 198, 219
food presentation, 187
olfactory, 185
European Convention for the Protection of Pet Animals, 261, 263, 274
euthanasia, 52, 76, 77, 92, 149

of cats in shelters, 120, 121, 123, 135
FCV, 191, 209, 211, 215
fear of cats, 60
fear-related behaviours, 95, 101, 198
FECV, 191, 209
feeding behaviours, 70, 99, 113, 238, 241, 242, 250
feelings, 24
feline calicivirus. *See* FCV
feline cognitive dysfunction, 113, 249
feline enteric coronavirus. *See* FECV
feline herpesvirus. *See* FHV
feline immunodeficiency virus. *See* FIV
feline leukaemia virus. *See* FeLV
feline lower urinary tract disease, 111, 112, 195, 216, 220, 229
feline odontoclastic resorptive lesions, 245
Feline Temperament Profile, 77, 192
Felis nigripes, 185
Felis silvestris
 catus, 1, 3, 228
 libyca, 1, 3
FeLV, 148, 151, 166, 167, 209, 210, 211
feral cats, 25, 80, 107, 121, 141
 bond with caretakers, 166
 definition, 142
 diet, 159, 164, 238
 effects of predation on wildlife, 121, 158, 161, 163. *See also* predation
 extinction of native species, 162
 health status, 169, 210, 219
 introduced species, 161

methods of control, 145, 148
  number, 145
  sources, 144, 157
FHV, 191, 209, 211, 215
FIV, 148, 151, 166, 167, 209, 210, 211
food resources, 1, 11, 27, 145
food selection, 250
  aversion, 240, 252
  moisture content, 241
  neophobia, 240
  palatability, 239, 241
  satiation, 241
Global Assessment Score, 31
glossary of genetic terms, 273
hiding behaviour, 30, 42, 101, 109, 182. *See also* fear-related behaviours
hierarchy, 12, 14, 27, 110, 198
home, 28, 65, 97, 101, 104, 185
home range, 6, 14, 15, 105, 197
house soiling. *See* elimination behaviour and marking behaviour
housing
  home, 197
  laboratory, 188
  licensing, 178
  shelters and catteries, 190
housing requirements, 179
  comfortable surfaces, 181
  conspecifics, 183
  contact with humans, 183, 191
  control over environment, 179
  nutritional environment, 187
  occupational environment, 187
  olfactory environment, 185
  quality of space, 181
  quantity of space, 180
  sensory environment, 185
  vertical dimension, 181

Human-Approach-Test. *See* behavioural testing
hunger, 97
hunting. *See* predation
hyperaesthesia syndrome, 112
hypothalamus, 26
identification of cats, 72, 157, 195
idiopathic cystitis, 38, 111. *See also* feline lower urinary tract disease
individual variation, 9, 29, 39, 40, 222, 240
infanticide, 6
infectious disease, 114, 180, 191, 205
  bacterial, 210
  boarding cattery, 218
  breeding cattery, 217
  carrier cats, 214
  horizontal transmission, 213
  impact on individual, group and population, 207
  methods of control, 215, 217
  multi-cat households, 217
  reduction of exposure, 212
  screening, 210
  shelter, 218
  vertical transmission, 213
inhibition of behaviour, 29, 31, 179, 222
keeping cats indoors, 78, 80, 98, 99, 103, 185, 194, 197
kill on site methods, 146. *See also* feral cats: methods of control
kittens
  development, 17, 61, 102, 239
  handling, 17, 61, 62, 101, 102
  nutrition. *See* nutrition
  post-meal behaviour, 62
laboratory, 25, 178, 184, 185, 189
LH, 27, 41

litter
  substrate, 105, 183
  tray, 104, 183, 198
luteinizing hormone. *See* LH
maintenance behaviours, 30, 179
marking behaviour, 94, 95, 97, 103, 106, 186
mating, 6
mental state, 24
miaow, 16, 69. *See also* vocal communication
motivation, 31, 97, 100, 103, 106
multi-cat households, 28, 97, 104, 106, 108, 109, 111, 198
neotenization, 54, 70
neuroendocrine response, 26
neutering
  costs, 125
  effects, 108
no-kill shelters, 77, 191
non-infectious disease, 205
  screening, 220
normal behaviours, 95, 196
nutrition, 250
  adult, 245
  geriatric, 246
  kitten, 243, 244
  obesity, 250
  oral health, 245
  queen, 246
  sick cat, 252, 253
nutritional problems, 237, 238
nutritional requirements
  anti-oxidants, 235
  carbohydrate, 232
  carnivorous specialization, 230, 250
  fat, 231
  minerals, 235
  protein, 230
  standards, 229
  vitamins, 232
  water, 243
obesity, 113, 195, 196, 229, 241, 250. *See also* nutrition
observational learning, 4
olfactory communication, 16, 96, 185
over-grooming, 38, 110
pain, 110, 112
  acute, 222
  chronic, 223
  treatment, 222, 223
parturition, 3
patterns of activity, 98, 184, 185
pedigree breeds, 64, 195, 196, 206
  characteristics, 64
  differences in behaviour, 64, 65
  welfare problems, 267
Persian, 64, 137, 206, 267
personality, 39, 62, 63, 65, 77, 179
pet cats, 25
pet foods, 229, 243
pet overpopulation, 119
  magnitude of cat overpopulation, 120
pet ownership
  benefits, 228
  cats, 48, 49, 56, 57
  children, 51
  dogs, 48, 50
  elderly, 57
  evolutionary aspects, 55
  family, 51, 59
  health benefits, 57
  pet death, 52
  prohibition by landlords, 131
  responsibilities, 71, 157
  risks, 72
  scale, 48
petting cats, 68
pheromone therapy, 107
play, 9, 18, 31, 98, 187

predation, 4, 97, 238
  behaviour, 97, 98, 242
  effects on wildlife, 78, 98, 160, 195, 197
  strategy, 238
preferred associates, 7
purr, 15, 16, 31, 70. *See also* vocal communication
quarantine
  and isolation, 216, 217, 218
  cattery, 28
rabies, 72, 121, 164, 209. *See also* zoonoses
reciprocal altruism, 3
relatedness, 2, 3, 7, 28, 107, 198
relinquishment to shelters
  characteristics of cats, 127
  characteristics of owners, 129, 131, 134
  risk factors, 76, 92, 93, 126, 127, 128, 131, 133
replacement, reduction and refinement, 188
rescue. *See* shelter
Resolution on Breeding and Animal Welfare, 266
Resolution on the Breeding of Pet Animals, 263, 265, 274
resting together, 11, 15. *See also* affiliative behaviours
retroviruses. *See* FeLV and FIV
road accidents, 78, 98, 196, 220
scratching, 15, 16, 38, 95, 96, 186, 198. *See also* marking behaviour
separation anxiety, 71
shelter, 25, 67, 76, 120, 121, 149, 178, 180, 185.
  role, 190
  single versus group housing, 192
Siamese, 64, 94, 137, 260

sociability index, 143
social
  behaviour, 18, 65, 66, 101, 107, 108, 180, 198, 199
  groups, 27, 107, 109, 114, 183, 192, 198
  learning, 4, 9
  organization, 2, 145
socialization
  sensitive period, 18, 61
  to cats, 18, 61, 193
  to humans, 18, 61, 62, 63, 100, 143, 193
stereotypy, 29
sterilization, 124, 125, 150, 158
stray cats, 25, 80, 137, 142, 143, 144, 157
stress, 28, 29, 31, 41, 42, 106, 109, 111, 182, 191, 192, 223
  and disease, 114, 215
  response, 26, 28, 39
submission, 12, 14
tail-up, 19, 70. *See also* visual signals
temperament. *See* personality
territoriality, 14
toileting. *See* elimination behaviour
touching noses, 7, 15. *See also* affiliative behaviours
toxoplasmosis, 72, 121, 165, 209. *See also* zoonoses
toys, 184, 187
trap and remove, 149. *See also* feral cats: methods of control
trap, neuter and return. *See also* feral cats: methods of control
  methods, 150, 151
  programs in Canada, 155
  programs in Germany, 156
  programs in Hong Kong, 157
  programs in Israel, 155

programs in Japan, 156
programs in Singapore, 156
programs in The Netherlands, 156
programs in the United States, 152
urination. *See* elimination behaviour
urine spraying, 16, 105. *See also* marking behaviour
vaccination, 207, 218
   effects, 207
   frequency, 209
   risks and reactions, 208
   vaccines available, 209
veterinary care, 71, 120, 123
visual signals, 14, 70, 96, 186
vocal communication, 15, 69, 222
vocalization. *See* vocal communication
weaning, 102, 244
welfare
   behavioural measures, 29
   interpretation of measures, 42
   physiological measures, 39
zoonoses, 72, 164, 166

Animal Welfare

3. N. Waran: *The Welfare of Horses*. 2002   ISBN 1-4020-0766-3
4. E. Kaliste: *The Welfare of Laboratory Animals*. 2004   ISBN 1-4020-2270-0
5. I. Rochlitz (ed.): *The Welfare of Cats*. 2005   ISBN 1-4020-3226-9

Printed in the United Kingdom
by Lightning Source UK Ltd.
126187UK00001B/41/A